THE SKEPTICAL SCENOGRAPHER:

ESSAYS ON THEATRICAL DESIGN AND HUMAN NATURE

For Deborah, who makes everything possible.

THE SKEPTICAL SCENOGRAPHER

Essays on Theatrical Design and Human Nature

Beeb Salzer

Foreword by
Edward Albee

Broadway Press
Shelter Island, NY
1995

Publisher's Cataloging in Publication
(Prepared by Quality Books Inc.)

Salzer, Beeb.
    The skeptical scenographer : essays on theatrical de-- sign and human nature / written by Beeb Salzer ; foreward by Edward Albee.
    p. cm.
    ISBN: 0-911747-36-2

    1. Theatre--Stage-setting and scenery. 2. Theatre-- Production and direction.   I. Title.
PN2091.S8.S35 1995   792'.025
                                       QBI94-21334

This book was set in Berkeley using desktop publishing hardware and software as follows: Macintosh Quadra 800, Microtek ScanMaker, Apple Laserwriter II NT with an Xante AccelaWriter 600 dpi controller, Adobe PhotoShop 3.0, and PageMaker 5.0. Illustrations were created in Fractal Painter. Book design by N. Deborah Hazlett and cover illustration by Beeb Salzer.

Manufactured in the United States of America; first printed February 1995. Printed on acid-free paper.

BROADWAY PRESS
12 West Thomas Sreet, P. O. Box 1037
Shelter Island, New York 11964
800-869-6372

FOREWORD

If you have assembled a book of essays, occasional pieces and ruminations covering a twenty-five year period, a reader must assume—must he not?—that you either still hold to the tenets and precepts assembled, or, if this is not the case, that you wish the reader to follow your mind from folly—say—to wisdom, unless you are a supreme ironist, in which case the trip might be seen to be from wisdom to folly.

These three would be interesting assemblings; a fourth—a sad one, but one comes across it frequently—would be a kind of "the hell with it; I thought it, so it might as well be included."

With Beeb Salzer, I think we can safely assume the first postulation to be the correct one, for this volume is of a piece—serious, stern sometimes, reasoned and (as time moves on) often wry and sadly funny.

As I was making notes for this brief foreword I put together a list of words I might want to use to describe Beeb Salzer's mind and method: honor, ethic, humanism, old-fashioned Liberalism, irony and humor were some I came up with, but with the awareness that in the fist of an Old Testament Prophet the last two—irony, humor—do not always clench comfortably with the first four—honor, ethic, humanism and old-fashioned Liberalism.

How strict he is with us—with himself!—as he postulates a moral/ethic code which will allow us to go about our daily business without self-disgust, which will make us no further worse than we are and may even make others better.

While he writes primarily about the theatre and his frequently uncomfortable place at its table, the hard lessons he has learned transcend the parochial and prove that we are, indeed, the sum of our parts.

I think I have had a happier time practicing theatre than Beeb has had. I have been—as a playwright—subjected less to the corrupt and corrupting infighting than he—as an artist/technician—

v

has been, and his argument is seldom with writers. (Bless him!) Oh, I have had run-ins with a few actors and actresses I wish were forcibly retired (or shot!), and a director or two has demonstrated an unwillingness to do exactly what I wanted exactly how and when I wanted it done, but, for the most part, I have found the theatre—audience taste and tastemakers aside—a rewarding environment in which to practice my rather sullen art.

Beeb Salzer—all his cavils aside, and one or two cavils of mine about his cavils—is one of those rare individuals who believes near-Utopian circumstance is still possible in the theatre—so long as we give up the selfishness of ego, work for the common good and, at the same time, maintain honor—and the more power to him for holding to that view.

Theatre is, at its best, a moral art—Liberal, humanist and ethical—and it is gratifying to know that there are some who will hold us to those standards.

Beeb Salzer's grapes are not sour, merely a little bruised.

Edward Albee
New York City
January 1995

PREFACE

· ●

In one way I have been lucky. So few theatrical designers and technicians write that those of us who do are like dogs that sing; it's not so much that we do it well as it's remarkable that we do it at all. On the negative side, when we publish in professional publications, there is almost no response. Hardly any letters to the editor praise or damn what we write. Writing for this particular audience feels to me like dropping a stone down a well so deep that no splash is heard. Occasionally, when I succumb to the temptation to drop bigger and bigger stones, I touch a raw nerve and a few outraged letters appear. Usually they point out excesses I have used, like a naughty child, to provoke reactions.

By contrast, the OpEd pieces I've written for newspapers always evoke responses. Sane people write letters to the editor. Others write to me at home. One man even had time to read the paper, find my home phone number, and call me, all before six in the morning. Those who track me down generally have special answers to all the problems of the day. One explained that the National Endowment for the Arts is in trouble because "they are fluoridating our water." Another assured me that the arts are not taught in the schools because "Jesus says not to."

On rare occasions, I have heard from readers of professional articles. A TV news camera man working for a national network in the White House wrote in response to "Working for the Boss," which appears in this collection. I had attempted what I supposed was a satire on a lighting effect which made Ronald Reagan appear heaven sent in his news conferences. The camera man said my article was not an exaggeration. They actually used the effect I described. Satire takes a beating these days.

In the late 1960s I started writing regularly because I couldn't design or paint for two years. In my late thirties I developed cataracts, a condition associated with old age. The accepted treatment for cataracts was a prolonged process during which the lenses "ripened" like grapes turning into raisins; no quick outpatient surgery and lens implants then. Left with little to occupy my time except for self pity, I started writing about what it was like for a designer to be half blind. "Confessions of a Blind Designer" was published in *Theatre Design and Technology* in 1974.

The old cliché that adversity prompts a re-examination of one's priorities proved true for me. I discovered that above all else, I am a person who needs to make things. It matters little if it is a painting, ground plan, piece of furniture, or an article; at the end of the day, satisfaction comes from a tangible product. I think it was John Updike who said that artists and writers share the same compulsion to make black marks on white paper. I also found that it is important to keep working despite difficulties. When I was first asked to design a "Madam Butterfly" for the Baltimore Opera, I begged off because of failing vision. Rosa Ponselle, the great soprano and then doyenne of that opera company, got on the phone and ordered me to do the sets. She was right. Perhaps most important with regard to this book, I realized that I wanted to express ideas that were impossible to communicate visually. I found myself arguing for the primacy of the text in theatre. As heretical as this may be for a designer to say, and as unfashionable as it may be in an age devoted to surface rather than substance, I believe in a theatre where ideas and language supersede spectacle and image.

In 1977, R. B. Heckler started *Lighting Dimensions*, a commercial publication for the entertainment lighting industry. I sent him a humorous piece which he printed. When he asked if I would be interested in doing a monthly cartoon, I countered with an offer to do a monthly column. We called the column "From the Balcony Rail." The articles continued when the Cardiff Publishing Company bought the magazine and when, a short time later, Fred Weller took it over. Fred became a friend, in part because he allowed me to write whatever I liked. Some articles were serious, such as reports of my visits to theatre conferences in Eastern Europe, others were outrageously silly. The late Lee Watson also wrote a column which, along with mine, insured readers of two extremely personal viewpoints each month. Both of our columns ended when Fred sold the magazine in 1986. Unfortunately there is now no journal in our field where an individual voice can be developed over time, where a jaundiced view of the profession is appreciated, or where we find some perspective on the value of the work we do.

Writing has never been easy for me. I've been helped by my wife who is my most critical reader. Over the years, the editors of *Theatre Design and Technology*, the late Ned Bowman, Tom Watson, and Eric and Cecelia Fielding have given me advice and support. Fred

Weller and his wife Susan made writing for *Lighting Dimensions* fun—more like a party than a chore. David Rodger, of Broadway Press, has made this book possible through his encouragement and hard work.

While I am thanking people, I must include both the people I have worked with professionally and my academic colleagues in New York and California They have provided the inspiration and the targets for these articles. There used to be a distinction between designers who taught and those who worked professionally. Now that almost all working designers have a university affiliation, there is a more healthy respect for academic programs. When I was finishing my graduate studies, my professors, Donald Oenslager, Frank Bevan, and Frank McMullan, suggested that I teach. Somewhat insulted, I did not heed their advice. I went to New York and worked in theatre, opera, and film for several years before deciding to combine professional and academic life. My professors were right—professors always are. Teaching has been a joy, a joy that increases year by year. I am grateful to my students, those who listened and believed me, and even more so, those who challenged me and made me constantly re-examine my beliefs. They planted the seeds that grew into many of these pieces.

I write to discover what I really feel or think. For me, the challenge is to employ a linear argument to reach a conclusion. In many of the pieces that follow, I was trying to understand the anomalous behavior of theatre people. Like psychiatrists who enter their profession because they are slightly nuts, I am intrigued by professional relationships because they do not come naturally to me. My explorations are fueled by my inability to automatically decipher the cruel, funny, generous, and imaginative things that theatre people do. I sometimes feel like a traveler in a strange land, reporting on the exotic customs of the natives. What a trip.

Beeb Salzer
San Diego State University
January 1995

CONTENTS

Education

Profession

DESIGN

[f. L. *designare* v.]

1. A plan or
scheme conceived
in the mind…

THREE CONCEPTS

First published in **Lighting Dimensions**: November/December 1982.

Thus the director is surrounded by images—some vague and merely felt and others definitely perceived in the mind's eye as literal pictures or in the emotions and thoughts as metaphors and symbols.

Frank McMullan in *The Directorial Image*

• • •

Hyman Kopf was in the lotus position.

"Hi, Hy."

He didn't answer. I tiptoed across his office floor which was slippery under toe with leaves and flower petals. The stale smell of incense lingered in the room. A futon cum sofa was piled against the wall. I sat in it and waited.

Kopf opened his eyes and saw me. "Salutations. Hope I haven't kept you long. Let's go to the tub and talk." He unwound his short round legs and, once upright, motioned me to follow him.

At the end of the hall we entered a steamy room. Twinkling sounds like wind chimes surrounded us. Kopf stripped off his white muslin top and trousers. I saw that I was to undress as well.

Sliding down into the warm water I noticed another person in the tub. "Meet my assistant, Peggy Incredible," Kopf said. Peggy winked and slipped lower in the water. Life is full of disappointments.

Kopf sighed and grunted, hissed and oohed as the water soothed him. (How much relaxing does a person need?) After a while he started to talk to me.

"You know that part near the end where he puts his eyes out? Well, I'd like to do that on stage. I see it as a kind of liberation. He's free from all the crap of society."

Peggy seconded with, "Yeah."

1

"You know, he's blind but can see for the first time. it's the first good moment for him. It's a happy climax to all that stuff laid on him about sleeping with his real old lady. I want the audience to see it as I do."

"Yeah."

"I see eagles soaring, leaves sprouting in spring, the smoke of an Indian peace pipe, a spider in its web…"

"Yeah."

"Rain falling on a dusty ball field, ants swarming on a Hershey bar wrapper, graduating midshipmen's hats in the air, a warm day in February, a cat's first mouse…"

"Super."

Kopf had turned to the side and Peggy was rubbing his back.

"A sunset after a storm at sea, a pig finding a truffle, wine running down a chin, warm sheets on a cold night, Kiss…quadraphonic on the desert"

"Beautiful, wow."

I could tell Peggy's pilot light had gone out a long time ago.

"I see butterflies coming out of his eye sockets," Kopf continued. "Can you do that? Huh? What a picture! Defeat—failure—disgrace and then an apotheosis: butterflies all around him. They lift him off the ground. He becomes a huge butterfly, spotted wings the color of blood and sun…"

"Yeah."

＊

The office on the thirty-second floor had a magnificent view. Through the waiting room window, on this clear day, I could time the intervals between takeoffs at the airport 17.35 miles across town. In Sixtus Byte's office, the shades were drawn against the view.

Byte himself looked as if he had shaved ten minutes before. His white shirt, gray pants, and short hair gave him the anonymous look of a corporate clerk. We sat in chrome tube and leather chairs and talked.

"I want to show you what I have done," he said in a soft, expressionless voice "The plan you gave me was a big help. I have plotted all the coordinates and assigned codes to the people. In simple terms, it allows me to see the show from anywhere in the house. Let me show you."

He lifted the top of the coffee table. The lid slid back revealing a console of buttons, lights, keys, and several CRTs, one with a light pencil, others with strange grids. Byte pressed a button. A screen appeared from behind curtains. He pressed a combination of buttons and the floor plan I had given him appeared on the wall screen.

"Notice the 'T' on the screen. When the blind seer comes in, he has his staff in his left hand. He tries to reach 'O' with his staff. If O is five feet six inches from center," Byte pointed to the O on the screen, "T can just miss him if he stands two feet nine inches on the other side of the center line. I have figured a six-foot staff and a two-foot three-inch arm length. O stands his ground and the staff passes right by his face. Look at it in elevation."

The screen blinked and there in elevation were the levels and stairs of the set. Byte punched the console several times and stick figures stood in the set. He pushed some sort of start button and the figures moved in jerky animation. O stood his ground while T swung his staff like a giant baseball bat.

"Let's look at it from the side." Byte made the setting turn on the screen so that we saw it as if we were sitting on the far aisle. T swung his staff but O was covering him so it looked like O had been hit. "Doesn't work that way, does it? Now the other side."

The set appeared on the screen from the opposite side and again the animation began. The staff still did not look right.

"You see, I need that lowest platform two feet four inches upstage. At that angle the blocking works." He made the set change on the screen and repeated the animation process. This time the business worked to his satisfaction and he pressed another button. In the corner of the room a machine began chattering. Byte walked to the machine, waited a few moments until it had finished its work, then pulled out a large sheet of paper. It was a printout of the set as changed in the last sequence.

"Here, take this with you for your modifications."

"But Sixtus," I had to ask him, "what if your actors don't fit the measurements that you have plotted out for the scene? What if T is taller and can reach farther? What if he hits O?"

"Not to worry. Look!" Byte punched up a list on the screen.

Male, 34 years, 3 months
    brown hair
    6 ft., 2 in.
    arm length 2 ft., 6 in.
Male, 34 years, 2 months
    brown hair
    height 6 ft., 2 in.

shoe 13
length of stride 35-1/2 in.
arm length extended over head 1 ft., 8 in.

At the top of the column of figures was the heading, "Casting Requirements."

~

The gilded letters "MIRAMOI" were worked through the rainbow that arched above the door. The entrance hall was lined with posters. Pin spots lit each one with what looked to be a kind of shuttered focus on a single portion of each poster. The brightest part of the poster was the line that said, "Conceived and Directed by Jose Pierre Miramoi."

The receptionist kept me waiting even though I knew J. P. was not busy. (Once I tried being an hour late for an appointment here and was still kept waiting the required twenty-two minutes. I have since learned that some people are on J. P.'s twenty two minute list and others on either a twelve or thirty-two minute list.)

At last, ushered into his office, I was greeted by Miramoi, who rose and stretched out his beringed hand with palm down. Was I meant to shake it or kiss it? The office itself always amused me. I had never seen one with mirrors on three walls and a closed circuit television focused on the desk chair. Both chair and the desk were on a low platform which raised J. P. just a few inches above everything else in the room.

The non-mirrored wall of the room looked like a theatrical pub, covered with autographed photos of people I recognized and many I didn't. Photographs, in fact, played a large part in Miramoi's life.

 For each of the past four years he had sent me an autographed photo of himself as a Christmas present. These gifts required elaborate thank you notes and my ability to express gratitude had worn thin. Even now as we exchanged greetings I noticed a pile of photos on his desk. No doubt he had spent the twenty-two minutes of my waiting time signing pictures in preparation for the holiday season.

"I have made some changes." He talked to me but looked at his own image in the monitor over my head. "My public would be disappointed if I did not include some things that they expect."

I nodded in agreement. He caught the motion in his peripheral vision.

"We will use the rear projection screen again at the end. After the chorus has their say about looking to the last day and counting no man happy until he has passed the final limit of his life, secure from pain, then we fade out the lights and use slides with the curtain calls. At the calls I want a picture of each character half the size of the stage as he takes his bow. We'll do the calls showing both the real actors and their blown up pictures at the same time. Then, after the individual calls, they will come back for the whole cast on stage."

"What slide do you want then?" I asked with as much innocence as I could muster.

"I really don't know," he lied. "Do you have any suggestions?"

I know a cue when I hear one. "If you wouldn't mind, I think your public would love a full screen picture of you behind the cast. It's only fitting, after all."

"That might be the thing. By the way, will you be around in two months when I do *Hamlet?*"

ART, NUDE CHILDREN AND PORNOGRAPHY

··●

First published in *San Diego Union/Tribune*: June 22, 1990.

On Tuesday a Cincinnati judge ruled that the Contemporary Art Center and its director must stand trial Sept. 24 on an obscenity charge involving the display of Robert Mapplethorpe photos of two children, one a young boy whose genitals are exposed, the other exposing a girl's genitals. The following piece responds to that decision.

• • •

I confess. I'm a child pornographer waiting for Sen. Jesse Helms, American Family Association's Donald Wildmon and Charles Keating to arrive at my door with an arrest warrant.

It happened this way. Twenty years ago my wife was posing for me in my studio. My three-year-old daughter was playing on the floor with her crayons. When my wife left to look after our younger child, my daughter said, "It's my turn, Daddy." She whipped off her clothes, climbed up on the stool, and struck a pose. I then did a three-minute sketch of a nude child.

My sketch does not have great artistic value, but I always felt it connected me to the tradition of artists painting nude children. Every museum with a collection of classic art has paintings and sculptures of nude babies.

Florence's Uffizi Gallery, as just one example, proudly displays paintings by Raphael, Masaccio, Veneziano, Signorelli, Ghirlandaio, Botticelli, Michelangelo, Titian, Parmigianino and many others—all of which show the genitals of little boys.

That the boy happens to be Jesus confuses the issue more. Are these great paintings obscene and sacrilegious? Or can nudity be condoned if the subject is religious? Will I be protected if I title my sketch "Infant Mary After Her Bath"?

It always seemed to me that naked babies were wondrously sensual but not at all erotic. Cuddling a warm, soft infant is one of the great pleasures of the world, and prompts feelings of love, protection and care.

In art, naked babies represent innocence and purity. Anyone who equates naked children with sexual lust has a real problem. I wonder about the mental health of those self-proclaimed guardians of our morality.

The Supreme Court has allowed every community to decide what is obscene. This further confuses me because I grew up in Cincinnati, worked for twenty years in New York and now live in California.

Each place has a different sense of what turns people on. In Cincinnati, I suspect even supermarket ads for cucumbers are a no-no. In New York, it seems that showing an inch of a woman's breast is not acceptable; in California it is.

Witness the full page ad for Madonna's "Blond Ambition World Tour" carried in the April 1 editions of both the *New York Times* and the *Los Angeles Times*. The ad shows Madonna sitting naked on a bed, glancing over her back toward us.

In the Los Angeles photo we see an inch or two of her breast between her arm and back, but for the New York reader this bit of flesh was air brushed out. Obviously, the citizens of the Big Apple are not supposed to know that Madonna has any apples at all.

Growing up in Cincinnati was a lesson in hypocrisy. Its citizens easily kept the city upright and clean because many of them simply crossed a mile or two over the river to Covington or Newport, KY. There, prostitution, gambling and other entertainments operated openly.

After I left, a young lawyer founded Citizens for Decency Through Law, an anti-smut group which made Cincinnati so clean it became impossible to buy *Playboy* or see an X-rated film.

The lawyer, Charles Keating, later moved West, continued his crusade as a Nixon appointee to the federal Commission on Obscenity and Pornography, and is now the leading villain in the Lincoln Savings and Loan scandal.

I find it curious that in our society, artists are vilified for expressing their understanding of a changing human condition while some of the religious and political leaders who are most protective of old fashioned family values are the ones caught with their pants down or with their hands in the till.

There is something wrong when we see more danger in naked babies than we do in naked greed. Sitting in my studio, I might feel less intimidated by our legal authorities if I had robbed a bank rather than sketched my little girl.

REGARDING CHAOS AND THE THEATRE
· ●

First published in *Theatre Design & Technology*: Spring 1990.

Consider a scenario: She became president of the United States because of a computer chip from Korea and an olive stone from Greece.

Her path turned toward the presidency years before when, as an aspiring actress on an opening night in Ohio, the computer chip in the lighting control failed, leaving her in murky gloom. Rattled, she went up in her lines. The local critic was devastatingly unforgiving because, at dinner that evening, he had broken a tooth on an olive pit. The actress was so demoralized by the reviews that she gave up acting, went to law school and 28 years later was elected president.

· · ·

If these events were purely physical phenomena, scientists would say they demonstrated the "Butterfly Effect," or in scientific terms, "sensitive dependence on initial conditions."[1] The image they use starts with a butterfly in Japan flapping its wings, moving a small amount of air and eventually causing a blizzard in Vermont several weeks later.[2]

The plays we produce often deal with such matters. We talk about fate or destiny and we debate whether we exercise free will or if we are subject to a cosmic determinism. Is the universe chaotic or is there a plan? We struggle to find answers, to find patterns and to find ways to exercise some control over our lives.

Scientists in the past decade have also become interested in occurrences which are chaotic. They are finding patterns in chaotic systems, and thus their discoveries are changing not only science, but concepts of fate and determinism, aesthetics, and human responses to chaos and order.

As with any new ideas, there are some scientists who discount the value of studying chaos. Nevertheless, the physics and math shelves in bookstores grow increasingly longer with volumes on chaos. For those of us with little background in math or science, evaluating the validity of the new theories must be left to others, since understanding the principles is hard enough. Even the new vocabulary takes effort. Such terms as fractals, stochastic, topology and strange attractors are impressive words to use at cocktail parties but they also represent concepts that can help us in technical theatre to make sense of the constant contradictions we deal with— the dynamic tension between order and chaos.

Scientists themselves define chaos in different ways. "Stochastic behavior occurring in a deterministic system."[3] "A kind of order without periodicity."[4] "Dynamics freed at last from the shackles of order and predictability...Systems liberated to randomly explore their every dynamical possibility...Exciting variety, richness of choice, a cornucopia of opportunity."[5] Whatever the definition, chaos research is being done in widely different areas of science from the study of ventricular fibrillation in medicine, to the turbulence of gasses and liquids in physics, to the formation of snowflakes in meteorology, to fractal geometry in math. (Fractal is a word coined by Benoit Mandelbrot to describe fractional dimensions which are a "way of measuring qualities that otherwise have no clear definition: the degree of roughness or brokenness or irregularity in an object."[6]

"...scientists are looking at chaotic systems as important pieces in the puzzle of existence."

Scientists who have been exploring chaos in nature have been thinking about phenomena which, because of their complexity, were formerly excluded from scientific inquiry or dismissed as aberrations in a simplified science of linear solutions. Now, with new discoveries, new thinking about what was previously unthinkable, and with the power of computers to iterate formulae beyond our previous imaginations, scientists are looking at chaotic systems as important pieces in the puzzle of existence.

Newtonian physics could not explain many phenomena. When we pour a cup of hot coffee, we can measure the coffee's temperature. We know that if we let the coffee stand, it will eventually reach room temperature. What we can't predict are all the different temperatures of the coffee as it cools and is subject to chaotic convection currents.[7] Similarly, Newtonian physics can chart the path of two moving particles, but the math for three particles is beyond our figuring. If three particles are impossible, it is useless to attempt to figure what the molecules in a gas are doing and how they behave when heated. However, we can measure pressures and temperatures and use a system of probability and statistics to predict and explain the physics involved.

Scientists using Newtonian principles were also forced to ignore slight discrepancies in complicated systems. What, after all, does a little friction matter? But it does, and even a decimal point or two in a problem run on a computer can produce unexpected and erratic results. Moreover, systems that appear simple enough to be fathomable are sometimes chaotic.

The Lorenzian waterwheel is an example. Picture a waterwheel made up of buckets with holes in their bottoms. If water is let into the buckets at the top at a rate that is less than what can run out of the holes in the bucket, the wheel does not turn. If the water is greater than what can run out, the buckets fill up and start to turn the wheel, continuing to lose water as they turn. When the water flow at the top is increased, the wheel picks up speed and the buckets reach the bottom of the wheel's circle still partly filled and start up the other side. As the wheel's speed increases, there is also less time for the buckets at the top to be filled with water and eventually the weight switches and the wheel changes direction. Edward Lorenz "discovered, over long periods, the spin can reverse itself many times, never settling down to a steady rate, and never repeating itself in any predictable pattern."[8]

Mathematicians are employing several new ways to look at the complexities of existence. One is called topology or "rubber sheet geometry."[9] Topology studies continuity, the gradual changes of shapes. Squares turn into circles and the trajectories of points plotted as two twisted planes intersect to form chaotic patterns. Trying to understand the concept of this geometry is an exercise guaranteed to challenge and expand the thinking of anyone working with visual design.

Another challenging idea for designers is based on the existence of many dimensions. To explain some theories, scientists posit that there might be ten dimensions, six of them tightly coiled but vibrating. And dimensions may not even be regarded as whole numbers. If a line is one dimension and a plane is two dimensions, a line that wriggles around to cover half of a plane can be considered to have one and a half dimensions.[10] Foreshadowing our difficulty in imagining ten dimensions, in the late 1800s Edwin Abbott wrote *Flatland*, which has become a cult classic. *Flatland* shows the difficulty faced by inhabitants of worlds having one, two or three dimensions in conceiving of worlds with more dimensions than their own.[11]

Benoit Mandelbrot's discoveries and theories go beyond stretching our abstract thinking. They speak directly to the aesthetics of natural and man made forms. His work with fractals not only gives shape to the visual representations of mathematical calculations, shapes startlingly beautiful in themselves, it forces us to reconsider the concept of what is beautiful.

"Why is geometry often described as cold and dry? One reason lies in its inability to describe the shape of a cloud, a mountain, a coastline or a tree. Clouds are not spheres, mountains are not cones, coastlines are not circles, and bark is not smooth, nor does lightning travel in a straight line…Nature exhibits not simply a higher degree but an altogether different level of complexity. The number of distinct scales of length of patterns is for all purposes infinite."[12]

The idea of infinite scale and pattern is explained in the literature on chaos by looking at the length of a coastline. If we measure the coastline of Britain on a map, one length is determined. A more detailed map will give a longer length. By walking the coast and measuring with a ruler, another figure is obtained. On hands and knees and using a microscope, still another measurement is produced and the process continues to infinity.

Coastlines are also examples of self-similar and scaleless forms. In other words, no matter at what scale or from what distance we look at a coastline, it looks like a coastline. From an airplane or when walking on the beach, the same characteristic bays and inlets are evident.

Mandelbrot found that computer images representing the formulae for irregular events in natural processes had a common and characteristic form, something like a gingerbread man with a spikey outline. Using the computer's ability to magnify any portion of the shape, he found that the spikey outline contained more gingerbread men and the same outline. More magnifications revealed that the complex outline continued to infinity. At any scale, the Mandelbrot Set is self-similar, the ultimate fractal.

In nature, although the patterns of self-similarity may only extend for three or four levels, they are all around us: in clouds, trees, leaves, blood vessels and feathers. The ubiquity of fractal geometry in natural forms leads to compelling theories about our perception of fractals and our innate sense of the beautiful.

CHAOS AND THE ARTS

 Mandelbrot argues that geometrical shapes such as those box-like buildings produced in the International or Bauhaus schools of architecture are not pleasing. Fractal shapes, combining order and disorder, complexity and self-similarity across scales, are so much a part of human biology that we find beauty in clouds, trees and seashores as well as in designs that contain these qualities. He cites the Paris Opera as an example. When we see the building at a distance, the complex outline is beautiful. As we get closer, the largest statues and decorations are pleasing. And when we are very close, there are moldings and details that continue to hold our attention.[13]

Science and postmodernists seem to have come up with the same ideas at the same time: Complexity is in, simplicity is passé. Although we have always judged works of art on the basis of their depth—classics reveal more meaning each time we re-experience them—are we now in an era which demands complexity of form as well as complexity of significance?

Fractal research has already reached the arts. Dr. Kenneth J. Hsu and his son Andrew published a paper which proposes that music—Bach is their example—can be reduced to a fractal framework. New notes added to the basic form will create "new" Bach pieces.[14]

As a sure sign that interest in fractals has arrived, a brochure from Fractal Generation in Miami, FL, announces "Fractal Artwear," a line of T-shirts and sweatshirts imprinted with fractal designs. As the brochure tells us, "Perhaps the most convincing argument in favor of the study of fractals is their sheer beauty." That beauty has been employed in Hollywood, where fractal computer art has helped produce otherworldly landscapes for movies. In 1984, an exhibition of fractal art was mounted and documented in a widely available book, "The Beauty of Fractals."[15]

Fractals and chaos suggest another aesthetic principle. They tell us that boundaries are the most interesting place to look. With fractals it is the place where just one more computer iteration causes the shape to change, with chaotic systems it is the instant between calm and turbulence, with nature it is the instant in which life is created.

In the theatre, we are thrilled by moments of recognition, by our own sudden insights, and by a character moving over the edge from one state to another. A character who is in love is not very touching, but the moment between not being in love and being in love, when the realization hits, can affect us deeply.

Designers have always been concerned with boundaries. They use moldings to cover joints and to add interest. They are aware of the line between negative and positive shapes, of the profile of walls against a cyc, of the shape of a costume and of the intersection of two colors. The added attention to boundaries may be like M. Joudain's realization that he had been speaking prose all his life, a perception that what has been done automatically has rules and reasons.

Artists have also been aware, on some level or another, of the effects of self-similarity, fractals and boundaries. The importance of chaos research is in the changes it is bringing to the way we think about the way the world works. Ilya Prigogine and Isabelle Stengers tell us, "Each great period of science has led to some model of nature. For classical science it was the clock; for nineteenth century science, the period of the Industrial Revolution, it was the engine running down. What will be the symbol for us?"[16]

Prigogine and Stengers answer their own question and suggest that the image for our time may be something like the dancing Shiva, "a junction between stillness and motion, time arrested and time passing."[17] Their choice of an image rests on the abandonment of the time employed in Newtonian physics in which phenomena were both linear and reversible, and where a machine-like universe was predictable—assuming we had all the information needed to calculate cause and effect. The concept of such a universe ignored much of nature that was chaotic.

Now that chaos is being considered, we must again look at the age-old problem of fate, of determinism versus free will. Progogine and Stengers offer us a theory. They see systems reaching a point where fluctuations within the system threaten its equilibrium. At this bifurcation point, chance operates within chaos and sends the system in a new direction, a path determined by chance. Determinism guides the system until a new imbalance occurs and the process repeats.[18] This theory has it both ways; both chance and determinism hold sway.

At this point, the scientific work being done on chaos affects theatre only tangentially. There are beautiful fractal graphics, aesthetic theories and a renewed debate about determinism. But on the simplest level, we might examine our attempts to eliminate chaos from our work and, like the scientists, search for a useful tension between order and creative chaos.

CHAOS IN THEATRE

In design and production, the Butterfly Effect has been our sworn enemy. Unforeseen glitches, gremlins and changes in concepts or personnel wreak havoc on our schedules, budgets and peace of mind. The definition of design is to create by planning ahead, to eliminate problems, to bring form out of disorder. Yet, try as we may, there are incalculable human and mechanical factors that thwart us at every turn. But while we battle on the side of order, thinking that chaos is the enemy, we tend to forget that art and creativity first rise out of the boiling caldron of chaos. All too often, by imposing budget and schedule restraints too early in the process, we are guilty of restricting the creative impulses of our colleagues. We have lost the balanced tension between order and chaos.

We created our own straight jacket. As non-profit regional theatre grew and as university theatres multiplied, a scheduled season of plays became necessary to insure cash flow and to guard against the perils of a hit/flop attendance pattern. A deficit-free theatre meant a subscription season which then meant planning so far in advance that creativity was circumscribed before it took its first breath.

Too often we are bound by a schedule of plays that demand design due dates even before directors really know what they intend to do. There is something amiss when scenery is finished before rehearsals begin and the shop, already working on the next production, is unable or unwilling to make changes. Something is wrong when rehearsal time is scheduled far in advance and each show must open to the public, ready or not.

"Something is wrong when rehearsal time is scheduled far in advance and each show must open to the public, ready or not."

These problems are often discussed at LORT meetings by artistic directors. Some theatres have tried selling vouchers which are exchanged for tickets when a production is ready, but for the most part, we labor in a system that has been forced on us by circumstance rather than by principle. It is therefore important that we do not accept these practices as the proper way to do things nor should we cease to look for alternatives. We must reclaim the artistic freedom that a lack of time and money have taken from us.

Metaphorically, the creative process is analogous to the difference between linear, clockwork, Newtonian equations and chaotic systems. Rollo May describes the way our brains seem to work on problems be they scientific or artistic.[19] Our conscious mind is the policeman of our thoughts and actions. It keeps us from saying everything we feel and, no matter how much we want to, prevents us from taking off our clothes on a warm spring day in the park. It is order, rationality, discipline, obedience and conformity. It guards, tames, judges and limits our ideas.

The unconscious part of our mind, the seat of fantasy and dreams, is chaotic, disordered, anarchical and irrational. It makes strange connections that are poetical and unexpected. Our creative ideas come from the unconscious and must somehow sneak past the gatekeeper of our conscious mind.

May says that we load the problem at hand into our mind and then it boils and mixes while we struggle to find answers. At some moment when our conscious thinking is not looking, has let down its guard, the solution bursts forth from the unconscious. That is why we so often get ideas while taking a shower or just before falling asleep. According to legend, Einstein had to use a dull razor because he got new ideas while shaving and often cut himself.

Moreover, great artists consciously flirt with chaos. In an intermission interview on a "Live From Lincoln Center" broadcast, Yo Yo Ma was asked if he thought about technique when he performed. His answer was, if memory serves, "Technique is the last thing I think about. A performance has to be on the edge of out-of-control."

The process of creativity and the willingness to take risks are, of course, only one side of an equation. An inspired painter attacks a canvas and then steps back to intellectually assess the work and a great cellist abandons thoughts of technique only after hours and hours of practice that concentrates on technique. The conscious critical mind makes choices between good and bad impulses created by the unconscious mind.

Audiences are affected by the emotional non-rational messages we send far more than they are by purely rational information. Metaphors are not logical. The better the metaphor, the deeper it probes into our gut-level emotional being. But an irrational metaphor is only valuable when it serves a rational purpose. The late John Hirsch used to say that he loved fireworks but couldn't watch them for long. The creative act takes place in the area of tension between emotion and intellect, order and chaos.

For an audience, great art is a reversal of the artist's process. The audience must permit the work past the conscious mind's wary sentinel into the dangerous depths of chaotic emotion where the artist started. Artists must be willing to take risks and audiences must be willing to take corresponding risks by opening themselves to an emotional experience.

Too often these days, we do not challenge the audience to take risks. We use our technical prowess such as complicated lighting and smoke machines to produce ersatz emotional effects. Our audiences are awed but not moved. We conjure up fireworks that are not metaphors. Rather than producing an emotional metaphor in the service of an idea, we cynically employ technology to create an effect for its own sake. Ironically, just as science is probing the chaotic, we theatre artists are using our new scientific tools in the service of productions that are manipulatively rational, uninspired and emotionless.

We often hear the excuse that audiences demand "lite" theatre. They can't be challenged. They have changed. They want sentiment rather than emotion, fireworks rather than metaphors. This may be true. Unhappily true. Most certainly, we see our society unwilling to find a dynamic balance between order and chaos. Today, chaos must be banished from our experience. Chance must play no part. Risk must be avoided. If there is an accident, someone must be sued and made responsible. Fate has no bearing on poverty or disease; victims are responsible for their own circumstances if they can find no one else to blame. There is even the idiotically naive concept mouthed by our young people: "I can be anything I want to be if I love myself enough."

It follows that in a society unwilling to marvel at the vagaries of fate, theatre audiences are unable to take risks or to give themselves up to a chaotic unknown. And audiences bereft of the concept that good fortune or ill-fortune are both dependent on unpredictable chance, are audiences incapable of exercising that emotion most necessary to the experience of drama, empathy.

THE ARTIST IN THE INSTITUTION

One of the social changes that has influenced our attitudes toward chaos is the growth of large institutions. The institutional mentality is not the only factor in our avoidance of chaos, but since so many of us now work in and for institutions, we should be aware of the conflict between creative chaos and organizational efficiency.

Institutional growth spawns bureaucracies, rules and regulations, forms to fill out, chains of command and standard operational procedures. Each tentacle of the organism propagates procedures intended to eliminate the necessity for making choices or facing unexpected situations. Chaos and bureaucracies are natural enemies, yet here we are, artists making use of fortuitous mistakes, chaotic inspiration, or even aleatory principles, caught in the web of a conformity based on convenience. The institution says we must follow the rules, while our art says we must break the rules.

Art and institutions have always been enemies. Artists are the first to be imprisoned by repressive governments and artists are the first to be denied tenure by universities which have rigid requirements based on traditional academic rules. Artists don't fit into an organizational mold and their differentness is not only an insult to the planning done by bureaucrats, it calls attention to the idiocy of unjustifiable restrictions on personal choice, reveals the folly of

17

inflexibility, and worst of all, ridicules the anti-human regulations imposed in the name of an efficient process.

Process is always at the end of a discussion of the tension between order and disorder. While professional theatres are dependent on the success of their product and search for an equilibrium that allows for creative chaos within the restrictions imposed by the perils of economic survival, educational theatres should stress process, with product a secondary consideration. Academic programs, however, are often more rigid in demanding compliance to organizational rules in an attempt to teach a "proper" process as the means to achieve a successful product. Maybe students cannot grasp a dozen different ways to do some thing. Or maybe teachers find it easier to evaluate student performance within proscribed limits. Some teachers may even be passing on, unquestioned, a process they have been taught. In any case, students deserve to be taught that process is not a frozen dogma of "how-to" rules, but is an evolving search for better ways. The tension between chaos and order is part of that search.

The collaborative process we use in a production, a microcosm of an organization dedicated to accomplishing a task, is a good example of this evolution and search. Each production evolves a somewhat different method of collaboration; theaters with permanent staffs differ less in the process from play to play than do productions with freelance participants. But in every production there is a jockeying for position and an implicit or explicit competition between individual artistic visions.

In our modern theater, this jockeying for self-expression has been limited by a hierarchy of power and a paradigmatic organization. With the text as a bible, the director is the anointed interpreter. The set designer, costume designer and lighting designer have input, in a decreasing order. Many of us were trained to honor this chain of command and have taught it as gospel because it was the path to a unified production. It has become a rule ready to be broken.

The feminist movement sees the rule bridling the creative process. It is hierarchical and thus modeled after patriarchal and militaristic constructs. Raynette Halvorsen Smith calls for a re-examination of the production process in light of new ideas derived from feminism, post modernism, deconstructionism and feminine performance art.[20] She does not, however, propose a new process, though a feminist process might substitute consensus for fiat and equality for hierarchy.[21]

The problem with such a structure, and possibly the reason Halvorsen Smith does not suggest it, is that consensus is a form of order more restrictive of creative chaos than patriarchal or matriarchal hierarchy. At least a dictatorial leader may take risks and mandate a personal and possibly chaotic vision. If we list recent new directions in theatre, innovations have come from individuals with strong personal visions: Grotowski, Robert Wilson, Peter Sellars or even the late Tadeusz Kantor, whose dictatorial power extended into the run of the play when he directed on stage at each performance. Are there corresponding examples of innovation by consensus?

When we consider collaboration, it is as a method of bringing order to the disparate work of many people. But what if collaboration meant a way of creating chaos? Director Robert Woodruff calls for a system in which there is opposition rather than agreement, many visions rather than one. He would like the designers not to talk to one another but each arrive at tech rehearsal with a personal expression of the play. He uses a musical image, comparing the process to a 16-track recording.

> So we lay out sixteen scores, and the mix comes when we're in the theatre doing a technical rehearsal or previews and seeing all the elements at once. Everybody may be playing all the time, and you have to make some adjustment to one or more of the scores. The work becomes more assaultive because it's denser, but it's not only the density of language. Not everybody has to play behind the text, support one melody. Each can play his own melody, and there doesn't always have to be agreement. There's a tonality that you look for, but it doesn't have to be harmonic. That creates an unsettling edge that contributes to live performance. It's not a closed experience, as in situations where you have a kind of roundness, a nice landscape that everyone agrees on and where the edges of it are taken away. When you have agreement it seals things. You've taken away the possibility of a dynamic that can come from disagreement.[22]

As a process, Woodruff's ideas come close to what the scientists working an chaos are doing. Having recognized that the world is not a neat, closed, linear, orderly system and that chaos is a major part of reality, they are looking for some meaning in disorder. The weakness in Woodruff's aesthetic is that he is subject to the Butterfly Effect. His sensitive dependence on initial conditions is his choice of collaborators. If his colleagues are out of tune, play wrong notes or don't understand the play they are scoring, no amount of sound mixing will stay disaster.

This may have been the case all the while. Great artists probably create order out of chaos, inventing their own systems while the most restrictive step-by-step process will never turn an incompetent into an artist. Chaos inspires the artist while rules are made to keep the mediocre from falling off the deep end.

ENDNOTES

1 James Gleick, *Chaos: Making a New Science* (New York: Penguin Books, 1988), 23.

2 Ian Stewart, "The Weather Factory," chap. 7 in *Does God Play Dice?* (n.p.: Blackwell, 1989).

3 Stewart, 17.

4 Gleick, 306.

5 Gleick, 306.

6 Gleick, 98.

7 Gleick, 24–25.

8 Gleick, 27.

9 Stewart, 63.

10 Stewart, 219.

11 Edwin A. Abbott, *Flatland*, 2nd ed. (New York: Dover Publications, 1884).

12 H. O. Peitgen and P. H. Richter, preface to *The Beauty of Fractals* (n.p.: Springer-Verlag, 1986); quote by Benoit Mandelbrot.

13 Gleick, 117.

14 Malcolm W. Browne, *New York Times*, 16 April 1991, sec. B, p. 1.

15 Peitgen and Richter, 12.

16 Ilya Prigogine and Isabelle Stengers, *Order Out Of Chaos* (New York: Bantam Books, 1984), 22.

17 Prigogine and Stengers, 23.

18 Prigogine and Stengers, preface by Alvin Toffler, xxiii.

19 Rollo May, chapters 2 and 3 in *The Courage To Create* (New York: W.W. Norton, 1975).

20 Raynette Halvorsen Smith, *Journal of Dramatic Theory and Criticism*, Spring 1990: 153–163; a paper presented to the Association for Theater in Higher Education, 1990.

21 Deborah Tannen, *You Just Don't Understand* (New York: William Morrow and Co., 1990), 119, 217.

22 Robert Woodruff, interview by Arthur Barstow in *The Director's Voice* (New York: Theater Communications Group, 1988), 313.

The Paradox Of Multi-Dimensional Scenographic Presentations Or Why Models Are No Damn Good

First published in **Lighting Dimensions**: December 1978.

More and more scenographers seem to be using models rather than renderings or sketches to demonstrate their scenic concepts. This trend is evident in recent exhibitions of scenography,[1] in interviews with scenographers,[2] and even in a recent book devoted solely to the process of model making.[3] The reason for the shift from two-dimensional to three-dimensional presentations is generally explained by showing a corresponding shift from two-dimensional stage scenery (wings, borders, painted drops) to three dimensional sets (units, thrusts, sculptured effects).

This explanation is plausible and is indeed a demonstration of positive reactions to changing esthetic patterns. There are, however, negative reasons for a greater use of models, and these, along with the larger artistic implications of designing with models, should be discussed before scenographers completely abandon the art of presentational renderings.

The very words "presentational renderings" have such an old-fashioned ring to them that to write them, not to mention defending their use, is an embarrassment in a "culture that has institutionalized change"[4] and has produced "the tradition of the new."[5] However, renderings have been around a long time and both their faults and advantages are known to, let us say, mature designers.

Among their advantages, not the least, is that renderings are a darn sight easier to carry on the subway. Certainly they can be done in a way that allows all sorts of scenic problems to be blended into a dramatic mist. Yet only a designer who cheats at solitaire persists in producing such sketches. But let us look at model making, for the advantages and disadvantages of that technique are still to be explored.

21

I suspect that one reason models are more in use is that many directors find them easier to understand. From experience, a scenographer finds which directors can, and which cannot, read plans and renderings. The disasters that result from a visually illiterate director, discovering that the set is not at all what he had expected, is said to be reduced considerably by use of a model. Even this safeguard is not foolproof. We all know that some directors can visualize a set from rough plans and sketches while others need the whole hand-holding treatment of renderings, plans, models, as well as a lengthy verbal explanation. Even so, it sometimes happens that a built and painted set on the stage is incomprehensible to a director.

Another reason for doing a model is that the designer is not very skilled at rendering techniques. Many young designers have not had enough training in perspective, drawing, or representational painting. I consider this an invalid reason for designing with models. Our concepts are always limited by our own technical prowess. To enclose them still further seems unwise. A writer does not use only two-syllable words because the longer ones are too hard to spell, nor does a musician play only in the key of C.

Larger issues become apparent when we look at the examples of settings produced by the design model technique. There are no statistics available (a thesis subject, perhaps) but I suspect that a high proportion of those productions in which the scenery dwarfs the actor were designed with models. (Perhaps I am prejudiced; I have not seen as many actual productions as I have seen photographs.) Nevertheless, each time I see a set, either on stage or in a photograph, that completely overpowers the human being on stage, I invariably find that the scenographer designed it with models. This relationship of cause and effect baffled me for some time.

 Logically, a three-dimensional model should show scale, and reveal the importance of a human figure against or in the scenery more clearly than a rendering. I wondered how a designer or director, with a scale model to work with, would fail to see that the scenery had taken an upper hand. I now believe that there are at least three ways by which the model misleads us in its representation of a built set.

Anyone who has ever presented a director a set model has noticed the "Doll House Effect." The cunning little toy brings out the child in the most serious director. Some can spend hours playing with the little figures. One director only uses one hand to move figures and furniture because with the other he sucks his thumb.

Of course, designers can also fall prey to the same temptations and herein lies the danger. The cuteness of the model, its charm, and its summons to a playful world of forgotten delights, acts as a kind of esthetic laughing gas, providing only joy and blinding us to the real problems of the full-scale set.

Another misleading characteristic of a model is that its scale is distorted when viewed by a full-size person. If we take the distance between the eyes of the full-scale viewer and translate that space into the reduced scale of the model, it would require that we have two persons sitting a couple of seats apart. We are thus getting a far greater three dimensional effect by looking at the model than we get in the theatre. Architects, recognizing this phenomenon can use a reverse periscope device which gives a monocular view and a reduced scale to a model.[6] Even these instruments are not completely satisfactory but they are an indication that architects acknowledge a problem where one exists.

A third misleading characteristic of models involves our cinematic perception of reality. All of us have had our thought processes and our way of looking at the world changed by films and television. In the theatre playwrights now write scenes that dissolve and fade from one to the other, and film is used as part of some theatrical productions. Cinematic thinking is different from theater thinking so that when designers and directors look at a set model in a filmic way, the model ceases to function as an analog of a real set. Writing many years ago, Béla Balázs asserted that film functions in the following ways that theatre can not.[7] Film can: 1) vary the distance between spectator and scene; 2) divide the scene into "shots"; 3) change angle, perspective, and focus within scenes and even within "shots"; 4) create a montage by assembling "shots" to make up a scene.

Most people look at a model in the manner a film camera would, moving around from one side to another, changing angles, getting closer or farther away, changing focus and perspective. I have never seen anyone sit still in front of a model the way a theatre goer sits watching a finished set. A model viewed in this cinematic way is as misleading as a rendering filled with haze.

Scenographers are in the business of solving visual problems. We should be able to overcome the shortcomings of viewing a model. In order to prevent the director from using the model as a toy, we must see to it that the director cannot touch the model. To guard against a false binocular scale, we must devise a way of viewing the model with only one eye. To eliminate the cinematic effect, we must limit the movement of anyone viewing the model.

The observant reader will immediately grasp that by ridding the model of its drawbacks, we have produced a visual presentation that has no moveable parts, is viewed as two-dimensional, and is seen from one fixed point. It is interesting to note that such a presentation bears a striking resemblance to a rendering. Obviously, more research must be done in this field.

ENDNOTES

1 See the catalog for *Contemporary Stage Design U.S.A.,* International Theatre Institute of the United States, distributed by the Wesleyan University Press, Middletown, Conn.

2 Examples appear constantly in periodicals or see several of the interviews in Lynn Pecktal's book, *Designing and Painting for the Theater* (New York: Holt, Rinehart, and Winston, 1975).

3 Darwin Ried Payne, *Materials and Craft of the Scenic Model* (Carbondale: Southern Illinois University Press, 1976).

4 Nicolas Calas, *Icons and Images of the Sixties* (New York: E. P. Dutton and Co., 1971).

5 A phrase attributed to the art critic, Harold Rosenberg. *Ibid.*

6 The catalog of Charrette of Cambridge and New York carried ads for this instrument called a "model scope."

7 Béla Balázs, *Theory of The Film* (n.d., reprinted, New York: Dover Publications, 1970).

THE SCENOGRAPHER AS AN ARTIST

First published in *Theatre Design & Technology*: Spring, Summer, Fall 1977.

In the mid-nineteen seventies the once unquestionable "verities" of contemporary esthetics have exploded like a balloon blown too full of air. All around, critics with needles and sharp pens in their hands are asking questions. It is my purpose in this, and the two following articles, to explore the position of the scenographer as an artist in our society and to ask some questions of my own. This first article deals with the Alienation of the scenographer. The second, titled Limitations, is concerned with the historical differences between the stage designer and the painter. And the third article, Communication, looks at the problems all contemporary artists face in speaking to a public.

ALIENATION

Alienation has several dictionary meanings. One is "not belonging, being different in nature, adverse." Another meaning is, "being separated from one's self, being insane." For the past ten years I have felt different, adverse, unable to accept an esthetic which seemed hollow to me yet was considered the proper direction for art in our time by those who gave the appearance of knowing. But now I have come to believe that the artists who paint monochromatic canvases, or who have themselves crucified or shot, or who move earth with bulldozers, or who simply pass out directions for moving objects on a wall are the ones who are alienated in both senses of the word.

As a scenographer, I cannot help feeling separated from an art world that rejects the very principles I need to function. This separation challenges me to examine my own principles in relation to the art I see exhibited and praised. I need to evaluate what I see in spite of warnings issued by critics that historically new ideas have always been laughed at and we cannot truly evaluate new art. They say that by opposing what we don't like or don't understand we risk being inhibitors of artistic progress. This reasoning is difficult to answer. For years I have felt left out and a bit paranoid. Only recently am I realizing that I have more company than I'd imagined.

I can date the beginning of my own alienation rather precisely. It was Tuesday, October 13, 1966 at 8:30 p.m. Until that time I had thought of myself as both a scenographer and a painter. I felt a part of the community of visual artists, even when painter friends would occasionally imply that my association with commercial theatre cast a shadow over my artistic integrity. And then came the evening (it was actually several evenings) in October 1966.

The ad in the paper said, "*Nine Evenings: Theatre and Engineering.* It's art, and engineering, and a little theatrics. It's important that you attend." It promised to be marvelous, perhaps a turning point in the arts. And how fitting that it was in the 25th Street Armory where the famed Armory Show of 1913 introduced modern art to America.

In actuality, the most dramatic thing at these events, as well as the only revelation of things to come, was the really short mini skirt, straight from England, worn by many of the girls in the audience. The performances were far less interesting. Steve Paxton, a dancer, devised an inflated tunnel, lined with loud speakers, through which the audience entered. He described his work this way:

> *This piece is a dance with a set. It is cast not only by those chosen as permanent population...but by those who have chosen to come and see it...With regards to air pressure and topography, this piece is not an airplane, is pretty much the opposite of an airplane, but much the rest of it is analogous.*[1]

Robert Rauschenberg's contribution, called *Open Score*, consisted of tennis rackets wired to broadcast a signal every time they hit a ball. The signals set off sounds and also lowered the lights in the Armory. When the lights were out, infrared lights and infrared television let every one see in the dark. Rauschenberg's program notes said: "Tennis is movement. Put in the context of theatre, it is a formal dance improvisation."[2]

Yvonne Rainer's piece, called *Carriage Discreteness*, featured the author directing her cast of "ten-odd performers" by means of walkie-talkies. Shades of Gordon Craig.

Other theatre designers and technicians I talked to were both bored and insulted by *Nine Evenings*. After all, theatre people have always felt responsible for entertaining an audience. To this end they have used electronic lighting and sound controls, hydraulics, and motors on a scale beyond the imagination of both the artists and engineers in *Nine Evenings*. It was like being asked by friends to watch the child of the house perform *Swan Lake*. A few minutes of this is cute, ten minutes get boring, and more becomes an absolute insult.

Now ten years later, I can understand why I was alienated by this attempted amalgam of all the arts and science. It is because the esthetic principles stated therein, either explicitly or implicitly, have since monopolized the art world of important galleries and influential critics. In part, the principles of this esthetic do not alter or add to the historical progression of man's search for beauty; it is an esthetic of rejection. It rejects all past theory. It rejects what is termed "extra esthetic elements", in other words, any meaning or message besides the physical existence of the work. It rejects the value of entertaining or interesting an audience. It rejects both skill in execution and skill in performance. And it rejects the search for beauty in a finished object or event in favor of concept or process.

I felt almost a part of the world again when I read this perceptive paragraph by Howard Bay:

> *A few words on Contemporary Art, as wisps and shards from the galleries now and then find their way on stage and lend that excruciating touch of chic to the performance. Painting, the application of paint to the canvas with a brush, wound up its history with deKooning. Since that day we have had various styles of decorative patterns and many objects that are really pieces of congealed art criticism. These products seem to be of vital interest to artists and those that sell art. These demonstration lectures in paint have been attenuated to the point of invisibility...I think we can say that the separation from human concerns in the Art Scene that matters is well-nigh complete.*[3]

Some time after Howard Bay wrote this, Tom Wolfe wrote *The Painted Word*, a long article (later turned into a book). The cover of *Harpers* touts the article, "Modern Art Reaches the Vanishing Point." A subtitle at the beginning of the article continues: "What you see is what they say." Wolfe's thesis, surrounded by some fuzzy art history, is that: "Modern Art has become completely literary: the paintings and other works exist only to illustrate the text."[4]

An immediate reaction to this article came in Hilton Kramer's next *New York Times* Sunday piece, "Signs of a New Conservatism in Taste." Kramer dismissed Wolfe's authority as an art historian but begrudgingly admitted: "...he has, alas, a very keen nose for news of our cultural life, for sudden shifts in loyalty and opinion, and a

"It was like being asked by friends to watch the child of the house perform Swan Lake. A few minutes of this is cute, ten minutes get boring, and more becomes an absolute insult."

27

ruthless grasp of the latest chic."[5] Kramer may see a change in taste or chic, but I believe the return to conservatism is a response to basic philosophical and psychological needs which are not being met. A trip to the Metropolitan Museum, or the Frick, quickly reveals our civilization's centuries old search for values more lasting than chic. I do not think I am being merely cantankerous when I compare the works of the past and the values they represent to what is shown in so many galleries today. I do not mean that contemporary artists must retain an interest in perspective or the representation of biblical scenes. But by rejecting all "extra esthetic elements" the artist's purpose becomes so private it disappears.

Howard Bay is neither inaccurate nor extreme when he complains of today's art reaching the point of invisibility. Consider Robert Irwin's exhibition which consisted of a piece of scrim hung across one end of the gallery, forming a simulated wall. (The scenographer makes simulated walls all the time and doesn't stop there.) According to John Russell in the *Times*:

> For some years now Robert Irwin has patrolled the frontiers of invisibility, never quite crossing them, but never quite turning his back on them either. His new piece at the Pace…suggests that Mr. Irwin has crossed these frontiers once and for all.[6]

This ironic put down is not strong enough if this show is in any way related to the total history of visual creativity.

I don't blame the artist alone for what has happened. Human ambition being what it is, he can quickly see that he must reject the past if he is to be shown, reviewed, and recognized. He gets the message in review after review, like the following by John Russell: "And we should know by now that one of the marks of good new art is that it does not in the least look like older art."[7] While Russell's statement is partially true, he has left out the fact that until recently, new art has been concerned with communicating themes common to mankind. What enrages Bay, Wolfe, and the public is that the contemporary artist has rejected a desire to communicate any theme of interest to others. As the host at a party who will not speak to his guests, he leaves his viewers out in the cold. This was precisely the case at a recent Guggenheim Museum show:

> It is the kind of exhibition that will be of great interest to the specialized public that keeps a close watch on the esthetics of abstract art. For others, however, the exhibition is likely to be a difficult, if not indeed a baffling experience…It does not, I think, offer the eye much in the way of visual pleasure. But it offers the mind—a mind that is well stocked with information about the development of abstract art since the late nineteen-fifties—a good deal to think about. For this reason, Mr. Marden will continue to be a favorite

*of the critics and the schools, and therefore, of the museums; and
the public, whether it gets much out of this specialized interest or
not, will have to resign itself to seeing a good deal more of his art—
and of his imitators—in the future.*[8]

As a scenographer, I am dependent upon every style and artistic concept from the caveman until now. I am heir to an esthetic legacy which allows me, to the best of my ability, to express myself to others. Why has the painter rejected his legacy and grown content with a kind of artistic onanism? Aldous Huxley, seer that he was, said in a letter in 1933:

It is deplorable that only bad painters should now undertake important and intrinsically significant subjects and that good ones should live in terror of all that is obviously beautiful, or dramatic, or sublime.[9]

The condition that Huxley deplored forty years ago has progressed to a ridiculous extreme today. It is hard for me to imagine a next step other than a return to some of the truths that scenographers have husbanded through this whole period.

In other artistic fields this seems to be happening. According to a cover story in the *Times Magazine*[10] Twyla Tharp is an inventive choreographer whose career is blossoming. Miss Tharp started choreographing in 1966, influenced by the very artists who presented *Nine Evenings*. Her style then was "minimalism," using nondancers. In 1967 she discovered what trained dancers could do and "she added dancing to her pieces."[11] In 1970 she began to use music and she allowed her dancers to be "performers." They discarded their earlier rule: "no trying to please an audience."[12] About this change Tharp is quoted as saying:

You can only keep this up for so long, it's self defeating. It's hard to like people who won't like themselves. I guess that we were afraid that by softening up and becoming...accessible, we'd be selling out.[13]

Miss Tharp's recent works for the Joffrey Ballet and the conservative American Ballet Theatre have been triumphs.

I chose to interpret Miss Tharp's journey as a victory for the principles that painters have rejected, and as a sign that there has been in Hilton Kramer's words, "a significant symptom of a shift in esthetic loyalties."[14]

If art is a reflection of our society, then amid all the complex forces at work we can begin to see why so many painters are alienated, insane. Our world seems out of control with internecine wars, revolutions, third world countries emerging with values and cultures different from our own, and governments saying one thing and doing another. The individual feels powerless and retreats into mystical contemplation or escapes through drugs. All the while, the artist withdraws by mumbling a monologue of interest to no one.

Although what I say may seem negative and ill-natured, I think my views are just the opposite. They spring from two sources. One is an optimism, about the nature of man and the future of humanistic expression.

It seems to me that our society is now searching for things to believe in. The vogue for nostalgia and the revival of old musicals reflects a kind of desperation to re-experience a time when everything was simple. The frantic excesses of the Bicentennial Celebration also cater to this need. But however much we know in our hearts that we neither live in the twenties nor during the American Revolution, and that the world of the past has gone forever, we seem to retain the desire for a commonalty in society. This seems to me the basis for our shift in esthetic direction. We can no longer be alienated in our arts or in our private lives because there are no answers to our deeper needs in catatonia.

The second source of my views is the theatre and my work as a scenographer. The nature of theatre demands communication with the public. The very exigencies of our work have limited us and thus protected us from those inward directed exercises which have alienated many painters from society. Therefore, the pressures of money and time as well as the need to please many people and to work with many people have not been deleterious to the scenographer. Rather, these pressures force him to be a social being who, today, more surely reflects the tastes of his time and the spirit of man than does the self involved artist.

ILLUMINATIONS

One of my first jobs when I came to New York was as ghost designer for a production by Salvador Dali and Abe Feder. Dali envisioned a ballet called *Gala* in which several men dance, then fight, and finally tear off one another's clothes, thereby revealing that one of them is a woman. She is hoisted onto a pedestal, whereupon milk spurts from her breasts and the men shower under this fountain vivant.

The set was several large framework cubes which were to be dunked into vats of some viscous liquid. In model form, Dali dipped four-inch wire cubes into a soap solution which formed beautiful bubble gems inside the cubes. When the cubes were enlarged for the stage however, a problem emerged: finding a viscous liquid capable of forming a bubble over a four foot space, a sixteen square foot area. After the largest chemical companies failed to find a solution, (use both meanings), Dali became impatient and took his project to Italy, where the ballet was performed minus the large scale bubbles.

The confrontations between Dali and the volatile Feder are still vivid to me but the lesson I learned was that a contemporary artist's thought process is not the same as a designer's. The designer is trained to solve practical problems in his search for an esthetic result; the artist either jumps immediately to the end product, as did Dali, or concentrates on process and forgets the end result. The artist, in either case, is expressing himself with a freedom the designer has never been able to exercise.

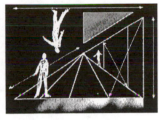

In purely esthetic terms, for the past hundred years the painter has been attempting to work in areas where it is impossible for the scenographer to follow. The painter's major intent in this period has been to destroy the effect of looking through a window and thus, to destroy the effect of observing the natural world from a distance that allows contemplation. The painter wants to produce a new relationship between time and space, to bring the viewer into an immediate and timeless experience of the work as a whole. To create this effect, painters have developed a series of styles beginning with cubism, going through abstract expressionism, and coming now to minimalism. Each style has become increasingly abstract.

The scenographer cannot hope to bring about an experience without time because the theatre is experienced within a framework of time. The scenographer can rarely follow the painter into abstraction because the human figure on stage is never abstract.

The painter was not always as free as he is today. In the Renaissance, artists were also designers. (Raphael was among those who designed for the theatre). Artists had their specialties but were often called upon to put their hands to a variety of design tasks. They designed buildings, table silver, monuments, servant's uniforms, and weapons. And even as painters they were faced with practical problems of decorating architectural spaces of a given size and of keeping within a budget by choosing a palette of affordable colors.

When perspective became a major interest of all visual artists, (see the work of Pierro Della Francesca) and the proscenium arch became a picture frame, the paintings of the time became stage designs. As Lee Simonson observed,

The vision of the artist as painter and the artist as scene designer were virtually identical.[15]

The designer's vision seems to have remained much as it was in the Renaissance. The painter on the other hand has undergone tremendous changes over the past four hundred years. Although the reasons for these changes are as complex as all the history of the intervening years, several theories have gained general acceptance.

One theory holds that oil paints freed the painter from large fresco commissions, thus allowing him to paint small, portable, private works. Another theory suggests that the disappearance of noble patrons freed the artist from working under commission. The development of photography is often cited as an agent of change because the camera has relieved the artist from the responsibility of creating a visual record of his world.

A historical explanation starts with the Industrial Revolution, which caused the breakdown of the master-apprentice relationship. With factories replacing craftsmen, young men left home to work in cities where factories were established. In the case of those whose talents led them to be artists, they no longer studied as assistants in the studio of a master artist, but now attended a new kind of school, the art academy. The academies produced far more artists than the ateliers had done because they were separated from the economic controls of supply and demand. In order to market the works of their graduates, the academies began to mount yearly shows.

The newly rich middle class bought the academy art but in the process established a standard of taste considered abominable by many artists. These artists rejected this taste and, as a reaction against it, developed the image of an artist that we recognize today. According to this image, the artist must oppose existing society and he must resist any attempt by the society to alter his own personal vision. The artist's honor depends upon his refusal to compromise his work for commercial gain.

But artists went further and dramatized their rebel status by adopting manners and dress that openly displayed their contempt for ordinary society and proclaimed their freedom from it.[16]

While the painter sought freedom from the middle class in order to create according to his own vision, the scenographer remained bound by the limitations of the theatre. As Richard Gilman says,

...the theatre has for several centuries been primarily a bourgeois art or enterprise and therefore a conservative one.[17]

In other words, the scenographer has been obliged to cater to the very audience which the painter has rejected.

However, it is not enough to say today that painters search for freedom, and the theatre has accepted the limitations of a bourgeoisie which is conservative. A recent commentator, Daniel Bell, has pointed out that the middle class itself, by losing its religious faith and its related work ethic, has itself turned to a search for self-expression and a desire to experience everything without limitations.

In the cry for the autonomy of the aesthetic, there arose the idea that experience in and of itself was the supreme value.[18]

Bell continues by claiming that a tension has been created because our cultural objectives of experience and release are in conflict with our economic objectives of rationality and efficiency.

There are several implications to these observations. First, if the artist and society now share cultural aims, the artist has lost a society to oppose, to push against. He is thus disarmed and his work lacks force.

Also, if Bell is correct, we have reached a point where the direction of either our cultural, economic, or political worlds, or all three, will change to relieve the tension that has developed. The shift to a more conservative taste in painting may well be a signal that we have reached such a turning point.

The theatre may also be at a turning point if it loses even its middle class audience which now prefers experiences which, because they rely on sensual perception rather than intellectual comprehension, are liberating in a more personal way.

Finally, the tension between an economy of regulations and a culture of independence is an analogue of the tension that exists for the scenographer. In the past, the artist working in a theatre that was a subsidized or commercial venture felt no pressure. But exactly as students today attempt to choose careers in which they can "do their own thing," theatre people have created a hierarchy of design jobs on the basis of how much control they have over the finished product. Theatre design is at the top of this hierarchy followed by film, TV, and commercials.

George Jenkins, who has worked extensively in theatre and film, recently described to me, with some dismay, how he had designed five complete proscenium arches to establish five theatres in a major Hollywood film. Even though they are skilled and practical artists, the director and cameraman used close-ups of the star in each of these five scenes and never once showed the arches, whose total cost equaled the cost of an entire Broadway show.

Even that form which is least limiting, the theatre, imposes tremendous limitations on the scenographer; the scenographer must please a producer, a director, and ultimately a public; he must work within a time limit with a given budget and a given work force; he must design for an existing space; and he must fit his designs to the meaning and spirit of an author's play. It seems to me that these limitations have very often created pressures which mirror the difficulties in our cultural life.

Given the limitations imposed on the designer from without, I find it fascinating that designers have added their own limitations and continue to preach the virtues of these self-imposed regulations. In 1941 Robert Edmond Jones wrote,

> *The designer creates an environment in which all noble emotions are possible. Then he retires. The actor enters. If the designer's work has been good, it disappears from our consciousness at that moment. We do not notice it any more.*[19]

and in 1975, Oliver Smith said,

> *I do believe that there are certain cardinal rules or guidelines which apply to all expert scene design. In other words, I think the purpose of the designer is to service the director and playwright and the composer, and not to show off as an easel painter. Therefore, I think the best sets are often the simplest, least fussy, and least distracting.*[20]

I don't believe that Inigo Jones, or the Bibienas, or Torelli, or thousands of other designers would agree with these statements. Possibly only those designers whose work spanned the years between Jones and Smith would agree in practice. But from Josef Svaboda to the designers of rock musicals, the greatest influences on design today come from those who ignore these admonitions. Therefore, Jones' and Smith's statements represent a historical phenomenon which does not hold true for contemporary taste.

Jones designed for a literary theatre best represented by Eugene O'Neill. For a variety of reasons, today's culture is not literary or rational. As Daniel Bell puts it, today's culture is,

> *…prodigal, promiscuous, dominated by an anti-rational anti-intellectual temper in which the self is taken as the touchstone of cultural judgments, and the effect on the self is the measure of the aesthetic worth of experience.*[21]

If contemporary culture is sensate, the ascendancy of rock light shows and the decline of the well-made play are easy to understand. One can see why opera, which is more sensual than intellectual, has gained a wider audience. Tom O'Horgan and Ken Russell have bombarded our senses while other directors mount classic Greek tragedies in which the actors make sounds but utter no words.

The scenographer is torn between contradictory tasks. On the one hand, as a visual artist, he has always represented the sensual side of theatre. But as a designer, limited by practical considerations, he has solved these problems intellectually. In the past he has limited the sensual impact of his own work so that it would not overpower the intellectual content of the author's work. But today the audience wants visual sensation and doesn't give a damn about intellectual understanding. And in a society that values "doing your own thing," the designer's ego demands that he consider himself a creative person or even an artist, yet he is bound by numerous practical demands which go contrary to the modern conception of creation and art. Furthermore, because most designers must design plays, operas, and musicals from every age of theatrical history, the designer faces a choice between two contradictory limitations. Either he serves as curator of museum pieces by designing with regard to the original period intent of the work or he designs within the limitation of pleasing a present day audience. In the process he may either destroy the play as written or he may lose his audience.

"The basic problem of the scenographer is that he is a Renaissance man, an artist, a mechanic, an inventor, a humanist."

The basic problem of the scenographer is that he is a Renaissance man, an artist, a mechanic, an inventor, a humanist. In short, he is an anachronism who must intelligently organize his many talents to function within imposed limitations. He is an anachronism in a culture which has no limits. Again, Daniel Bell:

> In the realm of art, on the level of aesthetic doctrine, few opposed the idea of boundless experiment, of unfettered freedom, of unconstrained sensibility, of impulse being superior to order, of imagination being immune to merely rational criticism.[22]

Many observers of the culture, Bell included, believe we have reached a turning point. The world of sensation has proved hollow. Irving Howe has suggested that our laissez-faire culture will be no more acceptable than was an Adam Smith laissez-faire economy.[23]

My own view is that the scenographer, by a strange reversal, is now the new avant-garde. By being four hundred years behind the times he will now be in the forefront of a culture which will return to art created within limitations. (Hang on to your narrow neckties; they are bound to come back into fashion.) For one thing, if new art must rebel against existing theory, the logical rebellion against an art without limitations is an art with limitations.

I would like to predict that scenographers will be the major influence on this new culture. But unfortunately, we will probably be ignored and be forced to watch as "artists" discover truths that we have known and practiced all along.

What we can do as scenographers is to abandon our defensive posture in which we claim to be craftsmen rather than artists. We must associate ourselves with those countless masters of the past whose great works were created within limitations. We must no longer be apologetic because our work contains ideas and feelings and because we communicate to a public.

Our greatest problem in this changing culture will be the same for us as it is for all artists, to find a message of importance and of universal interest to communicate to a fragmented society.

COMMUNICATION

Our theatre is in a transition, or crisis (depending on how we respond to events). Theatre has most recently been a form of verbal communication and is now becoming a theatre of visual and emotional communication.

While scenographers have always depended on non-verbal communication, the importance of words in the theatre has varied historically as each era has emphasized poetry, action, or spectacle. Words have also gained or lost importance when styles changed from realism to symbolism or to expressionism. There has been something like an inverse proportion in operation; the more important the words, the less need for scenery and spectacle.

Change is often painful, and for scenographers, the re-evaluation of long held beliefs is difficult on several counts. Many scenographers continue to believe that the essence of theatre is the Word. Also, some designers find that audiences today are so parochial in their experiences that they are unable to respond to some visual information. On the other hand, audiences can now be so international that they fail to identify the visual stereotypes recognized by a homogenous group. But more than anything else, as scenographers, we are troubled by the same things that are confounding all artists in this period; a fragmented society, a sense of hopelessness, and a lack of anything to say. Just at a time when all artists are suffering a communications vacuum, the playwrights and directors are turning to us and saying, "I don't have anything to say—you say it." And, of course, we have nothing to say either.

However, many social critics are expounding theories that directly affect our current concept of community and therefore explain the course of communication in the theatre.

It appears from what the experts are saying, that the conditions are not present which allow an artist to communicate to a public. There are no longer public traditions of meaning, nor a common faith, nor generally accepted myths or heroes, nor even a sense of what is real. And I believe that artists have indeed stopped communicating. They have settled, as has the public, for a culture of sensation rather than one of thought. The result is an art of the self, with no external standards, with no duty to, or interest in, the community, and thus no potential for communication.

The theories which prompt this view deal with ideas that at first seem unrelated, but subsequently fuse to give an overview of a troubled society.

One theme which recurs is the fragmentation of society. Even the units of society which were once cohesive and had a common purpose, such as a university community, have fragmented. According to the president of the University of Cincinnati there have been major changes in society and in his institution.

> ...we face a new movement of populism—the fragmentation of constituencies. On our campus we have more than 500 governance and interest groups, including a variety of women's groups, a gay lib, black organizations for students and for faculty members, and a faculty council for Jewish affairs. There is a loss of consensus, of community. It was Lyndon Johnson's tragedy to plead, "Come let us reason together" at a time when the various fragments scarcely even wanted to be together. The groups go their separate ways. They don't wish to be part of the mainstream of America.[24]

Many observers of our society have reached the same conclusion about the fragmentation of this society. In the arts, this has produced a communication vacuum. Art critic Harold Rosenberg writes:

> An art of objective reality untainted by parody would seem attainable only in styles shared by a society as a whole. The essential obstacle to an art of "feeling in the grand sense" is the social and psychological fragmentation reflected in the advanced art of the past hundred years, and this obstacle cannot be overcome by an intellectual act of force.[25]

This same critic reviewing a show of Jewish artists says:

> If the common condition of Jews, artists and non-artists alike, is an ambiguous, or dual identity, it is a condition that is experienced

uniquely by each individual, and cannot be expressed in common experience. Moreover, ambiguity of definition is a condition that is not confined to Jews, but is a general predicament—in this century of persons displaced from their class, their national, religious, and cultural heritage.[26]

Rosenberg's perception that a fragmentation of social units is connected with a loss of heritage and of identity, (which is cause and which is effect?) is a common notion. I think it is important for theatre people to consider this current phenomenon because our art is possibly the most dependent on historical continuity. From an audience's acceptance of theatrical conventions to the understanding of the stories of our plays, theatre survives on the connection between ourselves and the cultures of the past.

In an appreciation of Hannah Arendt's book, *The Human Condition*, Judith Shklar describes Miss Arendt's comparison of the classic Greek world, which was her ideal, to a contemporary society: She encapsulates society's and theatre's present predicament:

They (the Greeks) gave economic occupations a subordinate place; we worship them. Christianity at least preserved communal and contemplative life, but the natural sciences and systematic, self-oriented doubt destroyed these and far from liberating us delivered us into the clutches of necessity. Estranged from the world of nature and history we cling to mere life, unable to grasp reality and deprived of common sense, authority, and public traditions.[27]

One of the common theories put forward to explain our inability to grasp reality is explained by Maxine Greene, an educational philosopher. She quotes Daniel Bell, Hannah Arendt, George Steiner and David Hawkins in support of her contention that the most significant events of our time, being mathematical and scientific, are for the most part, beyond the understanding of a great majority of men.[28] Thus, society cannot even discuss those ideas which influence their lives. In fact, recent scientific discoveries involve particles so very small or events in the cosmos so far away that they are beyond the experience of man and therefore almost impossible to turn to poetic use.

Another group of observers sees our culture in the middle of a transition that is heading toward unknown territory. Alfred Kazin writes:

Technology lords it everywhere over the old-fashioned, art educated intellectuals brought up on modernism. Technology and bureaucracy, aided by all the sciences of manipulation and propaganda, mass culture and mass emotion, make up our public drama. Billions of people are being whipped into shape at any cost to the so-called individual, as in China; or are being left to die in the class wars, civil wars, merciless political wars of South East Asia; or are left to starve and die on the pavements of Calcutta and Bombay. Everywhere in the West there is panic, the most obvious slackening of will and hope and faith in anything more than the ever-more urgent whiplash of the money motive. We are under the gun.[29]

And Susan Sontag is equally grim.

...this civilization, already so overtaken by barbarism, is at an end, and nothing we do will put it back together again. So in the culture of transition out of which we can try to make sense, fighting off the twin afflictions of hyperesthesia and passivity, no position can be a comfortable one or should be complacently held.[30]

One further theory is important because it depends somewhat on all the others. Daniel Bell, in *The Cultural Contradictions of Capitalism*, explains that our society has turned to forms of immediate pleasure rather than retaining Capitalism's historical process of delayed gratification. And Richard Senneth's new book, entitled *The Fall of Public Man*, is reviewed in these words:

In mid-eighteenth century London and Paris...we even had separate modes of behavior for dealing with the public and private aspects of our lives. Away from our homes, we played public roles, dressed and spoke theatrically so as to project those roles and thus maintained a sense of civility. But in time, the rise of industrial capitalism and the ascendancy of a secular view of the world combined to undermine this public mode of behavior. The concerns of our private selves became transcendent. Hence, today we are steeped in narcissism.[31]

The result of narcissism on the arts is everywhere evident. It is even legitimized in a course given by The Institute for Architecture and Urban Studies, described in a mailing piece this way:

"Critical Issues of Art in the Seventies"

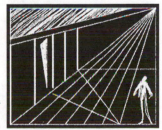

The art of the seventies seems to be generating an intense scrutiny of "forms of the Self." This manifests itself in not only the arts of sculpture and painting but also in film, photography, dance, and theatre. Strategies for exploring the self range from documentation through re-invented autobiography to the formal use of time.[32]

Self involvement is only one of several symptoms of our fragmented, disheartened, and transitional society. Maxine Greene says that artists today are communicating. She believes they are reflecting the general despair and the sense of being lost that we all have. That may be so. Private depression is a heavy burden when it lasts a whole generation and I believe we have turned to other forms as an antidote.

For the most part we have turned away from rational, lineal, intellectual, verbal forms of communication to more sensual forms. Bell's theory explains this switch while the attendance figures of ballet and opera support it. Legitimate theatre, the well-made play of ideas, has been deserted by the educated, affluent audience in favor of the more sensual entertainments of ballet and opera. Dance audiences in particular have grown each year so that New Yorkers can usually, at any given time, pick between several competing companies of excellence. The theatre-goer on the other hand, is lucky to find one or two plays of verbal excellence in any one season. Very often such plays are English imports and have limited runs because audiences will not support intellectual theatre, even with the most excellent casts.

There is one exception in today's theatre which I believe proves the rule. It used to be that a Jewish play in New York would find an immediate audience. Today, the Jewish play is being replaced by the Black play. It seems to me that theatre, needing an audience with common concerns, interests, and enthusiasm, is now able to communicate only to the one group in our society which still constitutes a community. The Jews have moved to the suburbs but the Blacks, because of external pressures, remain a geographical and psychological community able to share a theatrical experience as a group.

I imagine that unless there are drastic changes in our society, Black theatre will increase in prominence. It will be interesting and possibly sad to see what economic and sociological pressures do to Black theatre as it grows. There are already hints of difficulties in the form of exploitation films for Blacks, and of some Black theatre people who claim that the ordinary critical standards of theatre do not apply to them. There are stories of Black actors refusing to wear servant's costumes when playing the parts of servants. One hopes these are but growing pains.

Just as the written word, the printing press and the telephone have transformed our methods of communication and even man's thought processes, there are contemporary phenomena, while not as all-encompassing as the social theories cited earlier, which are, none the less, changing our culture's forms of communication. The use of drugs has contributed to private rather than shared experiences and has diminished areas of communication. I am speaking not only of the junkie nodding in a corner. He may be in a very dramatic private world. But as yet, no playwright has given a junkie wonderful things to say the way O'Neill did for drunks. Furthermore, I am concerned with what we will have to say to one another as society uses more and more drugs to alleviate physical and psychological pain. Many forms of mania, depression, and other mental problems are now treated with drugs such as lithium. While such treatment is of huge benefit to mankind, it does take away one whole area of communication. We may get to the point someday where a character in a play no longer need suffer anguish searching for the cause of his depression. His melancholy will have a physiological cause. A modern day Hamlet will not say, "to be or not to be..." He will say, "I'd better take a pill." I wonder if that will play?

Film and television have lessened the opportunity for communicating in the arts in two ways. First of all, our fragmented society has not been approached as separate interest groups with separate media offerings. Instead, commercial interests have demanded that as many people as possible be corralled into watching the same films and programs. The result is a public which has become accustomed to mindless entertainment which offends no one but which also demands no critical or intellectual effort by the viewer. Theatre is not for them. As Rollo May puts it, the need to reach a mass audience means that on television,

> ...originality, the breaking of frontiers, the radical newness of ideas and images are at best dubious and at worst totally unacceptable.[33]

The second thing the mass media has done is to give us instant replay reality in our living rooms. In doing this, crucial events in our history are denied the chance to become mythic by being retold artfully. We will never again have a Homer telling of Odysseus because our own epic stories, like the Kennedy assassinations, are recorded, frozen as they happened. Compared to the real event, any dramatization, is somehow wrong, off key, tame. When a dramatization of recent history is attempted, such as, *All The President's Men*, the whole focus of classic theatre is reversed. The main characters are off stage and we see only the messengers, the minor players. What would *Oedipus* be like if he never appeared on stage? We know what *Hamlet* would be like because Tom Stoppard has shown us.

A vacuum must be filled. We search for messages to fill our communication vacuum. It seems that no message is too odd to be given a hearing. In terms of personal and psychic fulfillment, there is a growing list of programs seeking to fill a vacuum of meaning in our society: est, Esalen, Rolfing, Transcendental Meditation, Transactional Analysis, group analysis of many sorts, Zen, Yoga, Moonies, and many other sub groups and combinations of theories. In the arts there is also a wide range of very strange performances trying to find an audience. For some time nudity and pornography have sought to satisfy what should be a universal interest. Because doing is more interesting than watching, theatrical sex has become more and more kinky. We are also treated to freak shows in which the line between performance and reality is blurred. The singer, Patti Smith, has become popular with an act reviewed in these words:

> The word "insanity" may seem a little strong; this listener hasn't been inside Miss Smith's head. But she acts crazy sometimes, and if it's an act, it's an act she plays so intensely that it's become, its own kind of reality…It was a performance "terrifying" in its intensity…At one point during one of her songs she slumped to the floor and started banging her head against the pipe organ.[34]

Excesses in the arts have prompted many critics to predict that we are in a period of decadence comparable to the worst of the Roman Empire, with a similar fate in store. But there is also a move away from this path. Articles are now appearing which claim that we are returning to both political and esthetic conservatism. A lead article in the *New York Times* "Arts and Leisure" section has a headline which announces: "A Yearning for 'Normalcy'—The Current Backlash in the Arts." The author, art critic Hilton Kramer, goes on to say:

> The truth is, much of what has passed for being avant-garde in recent years has proved to be extremely boring and extremely trivial—a mere charade of the great age of experiment in the arts, and of no great esthetic merit in itself…The taste now is for clarity and coherence, for the beautiful and recognizable, for narrative, melody, pathos, glamour, romance, and the instantly comprehensible, for empathy rather than entropy—for art that is a pleasure rather than a moral contest.[35]

All of these entertaining qualities that Mr. Kramer lists are evident in the public's current interest in nostalgia, in the bicentennial, and in numerous revivals now on Broadway. But unfortunately, we cannot survive on the messages of another era. The revivals may bring a breathing space from the pounding sensations of current culture but they carry no relevant communication. Alfred Kazin says:

> What are we nostalgic for? Why so many commemorations, appeals to the past, imitations of what has never really disappeared from our lives? Nostalgia is never for what we have lost, but for that part of ourselves, like childhood, that we no longer understand. Nostalgia is the effort of the too-conscious present to understand itself in the light of the unconscious past. Nostalgia is the effort to rejoin ourselves to ourselves—when the effort is in one sense, too late.[36]

Another, but equally, negative explanation of nostalgia comes from Jean Shepherd.

I think most nostalgia is a sickness. It's symptomatic of a deep cleavage in American life. It's one of the only things Americans have left in common—the past. How else can a kid from Hammond, Indiana and a kid from Brooklyn talk to each other? They have to talk about Bogart because they don't understand anything else about each other's lives.[37]

So we can't go home again. As theatre designers, our stock-in-trade clichés, stereotypes, and prejudices have been taken from us. The man with the long hair is not a hippie, he's a candidate for president; the mad man is not pitiful, he's a rock star earning a million dollars a show; and the babe in the slinky, low-cut, red dress is not a whore, she's your grandmother. We remain in a communication vacuum.

Finally, the point of all this is to explain why the absence of significant verbal communication has had a profound effect on the course of scenography in the past few years. The most published and probably most imitated scenographers today have abandoned the theories of design which three generations of designers in the United States have accepted as dogma. That dogma, according to Robert Edmond Jones, was:

A setting should not be a thing to look at in itself. It can, of course, be made so powerful, so expressive, so dramatic, that actors have nothing to do after the curtain rises but to embroider variations on the theme the scene has already given away.[38]

And in Prague, in 1976, Jo Mielziner repeats:

…the designer must be the servant of the director. Designers should not be leaders, they should be collaborators.[39]

But these statements depend upon a theatre of verbal ideas in which overpowering visual effects would kill the intention of the play.

Compare Jones' and Mielziner's words with these of Ralph Koltai, the English scenographer:

I'm interested basically in finding a concept for a production. I tend to work with directors who like me to find the concept. And I tend not to work with directors who have very strong ideas themselves.[40]

And Robin Wagner speaking of Tom O'Horgan:

He's very open and free about using anything that he feels is good. So if you come up with a seemingly crazy idea and it works for him, he wants it and you use it. He has absolutely no ego that way…someone had made a comment in an interview that I had a tendency to over-design and Tom's comment was, "No one can over-design for me."[41]

Or an article in this Journal which said:

I think that the evidence suggests that Svoboda's work in the 1970's shows relatively more of the traditional designer than of the scenographer as he himself has embodied that term...I suggest that the cause is closely related to his lack of work with a certain kind of director, one with whom he can interact fully and freely and almost intuitively...[42]

From these statements and from any number of sets we have all seen, it is clear that we have crossed a frontier in theatre. We are no longer completely bound by the old rules of scenic restraint. As scenographers we are even being prodded forward by Clive Barnes who has said that "Broadway design is scenically locked into the 1930s." He continues,

certainly we have never seen the sort of total visual concept that can occasionally be seen in opera or dance. Wake up, Broadway! There are images out there to be captured. They are not using gaslight anymore.[43]

What Mr. Barnes fails to recognize is that a conflict exists between the old theatre of ideas, of words, of humanistic principles, and the theatre of the senses, of visual supremacy, of mechanistic magic.

One spokesman for verbal communication is Barnes' co-critic on the *Times*, Walter Kerr, who writes:

What is very clear is that words, those strangely supercharged constructs by means of which a sustained intellectual intensity is generated, are not only the original tool of the theatre they are the only remaining tool of the theatre, the one means the stage has of coping with—maybe even whipping—the competition. The stage can't compete visually; it's got to compete verbally.[44]

"...must we jazz up the classics to interest such an audience, and thus destroy the true value of the play?"

If, as scenographers, we have only two choices; either to use restraint and be servants of the word or take command and provide visual fireworks, then our position is relatively simple. However, each production we design forces us into aesthetic decisions which are complicated by the state of our culture. Given a society, accustomed to mindless TV and sensual self-fulfillment, can we, as scenographers, offer this society productions of classic plays which stress beautiful language? Or must we jazz up the classics to interest such an audience, and thus destroy the true value of the play? On the other hand if we choose to make the visual production of utmost importance, even in a new play, what message of significance can we, as scenographers, transmit? For me, the huge, and often beautiful sets of designers, like Svoboda, are by their scale and power anti-human, and thus corroborate Maxine Greene's contention that today's art communicates the powerlessness and despair of man. In this case the unimportance of the character, dwarfed by his surroundings, serves as an additional defeat of man in his struggle against the forces of irrationality, of bigness, and of technology.

I am not an enemy of innovative and visually powerful scenography. Alwin Nikolais's work exhibits great technical skill and power, yet it is employed in the service of a traditional humanism. His advantage lies in the fact that he is choreographer, composer, and scenographer to his own works and that he is working in dance which has never been a verbal form of communication. Rather than unbalancing a work by causing one element of a production to take precedence, his work gains power through its unity and its use of technology. His art is always scaled to the size of the human body.

Few of us are fortunate enough to be playwright, director, and scenographer all at once. We work in theatre where responsibilities are fragmented. We live in a society that is fragmented, that is in transition, that is becoming more and more sensual, and that cannot even discuss the major events of its time. As scenographers, we have entered a permissive era where the old rules no longer operate. We are often asked to design classic plays and operas and in doing so are expected to use an aesthetic which has no relation to the original. As artists, we are expected to communicate and the only universal message today is despair. What are we to do?

I wish I really did have an answer that I could pronounce with conviction. But I see several directions and none of them are satisfactory. We can become museum curators attempting to foster a new appreciation for verbal theatre. We would have a hard time finding an audience. Or we can throw off the traces and work with directors who permit us to mount spectacular and sensual productions. But when we do this we seem to exclude vast areas of things that the theatre is able to say. Or, we can work intuitively, never considering the philosophical implications of what we do. If we do this, we are adrift in a meaningless activity that has lost the continuity of both history and the future. Or we can become so engrossed in technology, new lamps, new dimmers, new plastics, that like much of our culture we substitute things for what little humanity there still is in our work. It seems criminal that of all people, those in the theatre should take this course.

It is important for scenographers to realize that their new found importance and power in the theatre stems from the kind of messages they are capable of transmitting. Technical theatre can command sensual and emotional messages such as fear, excitement, calm, mystery, or even sexual desire. But we cannot talk of loyalty, honor, generosity, duplicity, or greed, but then, no one seems to be discussing such things any more.

I am concerned because intellectual quests rather than sensual satisfaction have raised our civilization to whatever heights it has reached. Now that we seem to distrust our own culture and the rationalism and humanism on which it rests, I fear that we will swing too far the other way in an attempt to explore our neglected emotional and physical selves.

As audiences, and critics, such as Clive Barnes, demand more sensual productions, we, as scenographers, must recognize that our growing importance in the scheme of things comes at the expense of a theatre of ideas. For me, this is a pyrrhic victory.

ENDNOTES

1   Program for *Nine Evenings: Theatre and Engineering*, 25th Street Armory, New York City, October 1966.
2   Ibid.
3   Howard Bay, *Stage Design* (New York: Drama Book Specialists, 1974).
4   Tom Wolfe, "The Painted Word," *Harper's*, April 1975, p. 57.
5   Hilton Kramer, "Signs of a New Conservatism in Taste," *New York Times*, 30 March 1975, sec. D, p. 31.
6   John Russell, *New York Times*, 21 December 1974, p. 22.
7   John Russell, *New York Times*, 27 December 1975.
8   Hilton Kramer, *New York Times*, 15 March 1975.
9   Sybille Bedford, *Aldous Huxley: A Biography* (New York: Knapp, Harper and Rowe, 1974).
10 Deborah Jowitt, "Twyla Tharp's New Kick," *New York Times Magazine*, 4 January 1976.
11 Ibid.
12 Ibid.
13 Ibid.
14 Hilton Kramer, "Signs of a New Conservatism in Taste," *New York Times*, 30 March 1975, sec. D, p. 31.
15 Lee Simonson, *The Art of Scenic Design* (New York: Harper and Brothers, 1950).
16 Several art historians present the same theory. It can be traced through several chapters of E. H. Gombrich's *The Story of Art* (New York: Phaidon Press, 1957; distributed by Garden City Books).

17 Richard Gilman, *The Making of Modern Drama* (New York: Farrar, Straus and Giroux, 1974).

18 Daniel Bell, *The Cultural Contradictions of Capitalism* (New York: Basic Books, 1976).

19 Robert Edmond Jones, *The Dramatic Imagination* (New York: Theatre Arts Books, 1941).

20 Oliver Smith, interview by Lynn Pecktal in *Designing and Painting for the Theatre* (New York: Holt, Rinehart and Winston, 1975).

21 Daniel Bell, op. cit.

22 Ibid.

23 Irvin Howe, interview by John Russell in *New York Times*, 16 February 1976.

24 Warren Bennis, in a speech before the Cincinnati Historical Society, reprinted in the Yale Alumni *Magazine*.

25 Harold Rosenberg, *New Yorker*, 12 May 1975.

26 Harold Rosenberg, *New Yorker*, 22 December 1975.

27 Judith Shklar, *New Republic*, 27 December 1975.

28 Maxine Greene, *Teacher as Stranger* (Belmont, CA: Wadsworth Publishing Co., 1973), 105.

29 Alfred Kazin, *New Republic*, 17 January 1976.

30 Susan Sontag, "Notes on Art, Sex and Politics," *New York Times*, 8 February 1976.

31 Christopher Lehmann-Haupt, review of *The Fall of Public Man*, by Richard Sennett, *New York Times*, 6 January 1977.

32 Description of a course offered by the Institute for Architecture and Urban Studies, New York City, Rosalind Krauss, course moderator, Spring 1976.

33 Rollo May, The *Courage to Create* (New York: W. W. Norton & Co., 1975).

34 John Rockwell, review of Patti Smith, *New York Times*, 28 December 1975.

35 Hilton Kramer, *New York Times*, 23 May 1976.

36 Alfred Kazin, *New Republic*, 17 January 1976.

37 Jean Shepherd, "Bridging the Gulf Between Indiana and Brooklyn," interview by Martin Jackson in *New York Times*, 19 December 1976, sec. D, p. 33.

38 Robert Edmond Jones, *The Dramatic Imagination* (New York: Theatre Arts Books, 1941), 26.

39 Jo Mielziner, interviewed by Iva Drapolova in "An Interview with Four Leading American Designers," *Theatre Design and Technology*, Fall 1976, p. 45.

40 Ralph Koltai, interviewed by Glenn Loney in *Theatre Crafts*, January/February 1977, p. 65.

41 Robin Wagner, interviewed by Lynn Pecktal in *Designing and Painting for the Theatre* (New York: Holt, Rinehart and Winston, 1975), 390.

42 Jarka M. Burian, "A Scenographer's Work: Josef Svoboda's Designs, 1971–75," *Theatre Design and Technology*, Summer 1976, p. 33.

43 Clive Barnes, *New York Times*, 8 December 1976, sec. C, p. 19.

44 Walter Kerr, *New York Times*, 9 January 1977, sec. D, p. 21.

Take A Meeting

First published in *Lighting Dimensions*: March 1982.

"I'm happy that you all got here today. We're all busy so we'll make this as short as possible, but I did want to get you all together so that everyone knows what is going on. We'll get together again when it's necessary. Do you all know each other? Let's see, I'll start here—Harry, Gloria, Pierre, Alvin, Jaynie, Christopher, Mac, John."

"Excuse me, Manny, before you get into something else, can everyone here get me a bio as soon as possible, we are going to start…"

"We'll get to that in a minute; first I want Harry to take over. I know he has a list of things he wants to discuss."

"Thanks, Manny. First, let me say that it is good working with you. Those of you who have worked with me before know how easy I am to get along with. Sometimes I have to say no to things you want, but remember that I'm on your side. I'm looking forward to working with those of you I don't know, and I'm sure this is going to be a pleasant experience. I'd like to start with Pierre telling us what his concept is and what this is going to look like."

"Oh, Harry. Do I really have to go first? Alright, if I must. Well, you see, when Joshua and I started to talk about this, he had some really thrilling ideas about the show—not experimental enough to empty the house, but new ideas that are really quite wonderful. I wish Joshua were here to tell you all this. So many are his ideas, and he does express himself so well. Anyhow, what we want to do is get rid of all that gloomy stuff and make this fun. This is a show about power. Of course, sometimes we have to be serious, but by having light moments we only make our serious points more effective."

"Pierre, can you give some examples?"

"Gee, Christopher, I'm glad you asked, I do tend to ramble. OK, take that scene when she comes down the stairs washing her hands. We're treating it like a dream sequence. The stairs are plastic and back-lit. Jaynie can do wonderful things with color in there. On either side of the stairs are chorus, just the girls, I think holding bowls—you know, for washing in. Except that the bowls are filled with confetti—red confetti. As she goes down the stairs, the girls toss the confetti like rose petals. It is a really strong visual statement and, you know, it's a very original concept."

"Pierre, excuse me. How do I get the confetti off the stage for the next scene?"

"Well, Christopher, the next scene has all those soldiers. And the ones in the back will have brooms attached to their pikes. Sweepie, sweepie—the confetti is gone."

"What do you see the girls wearing in this?"

"That's up to you, of course, Alvin. But the scene is basically pink and brown. And since they have all just gotten out of bed, I sort of saw some kind of medieval baby doll nighties. I can see, John, that you are worried about those bowls."

"That's OK, Pierre, I saw some beauties at Bloomies last week. Silver things. I'll show you. All I'd have to do is take Bill Blass's name off them."

"John, remember, there are 36 chorus girls. Can't you fake it with sprayed Tupperware?"

"Come on, Harry, every show you give me the same routine."

"Yes, and every show you are over budget."

"And every show it's your budget and too low."

"OK, fellas."

"While we're stopped, can I ask everyone to get me a bio as soon…"

"Gloria, we'll get to that. Mac, do you see any problem in a plastic stair?"

"Just have Pierre leave me some room to support it. It can't be all translucent. It has to stand somehow. I remember a stair that I made for Belasco, the old man, 30 risers, double curved, no visible support, 42 chorus girls in lead costumes bouncing up and down it, covered in gold leaf…"

"Thanks, Mac."

"…had to use double dove tails in split…"

"Thanks, Mac. Pierre, you want to continue?"

"Sure, Manny. Then there's the big banquet scene where the ghost appears. We're going to have this six-piece combo on stage playing dinner music. We kind of like the idea of the ghost coming out of the band. Some kind of magic, you know. Alvin, what do you think?"

50

"Off the top of my head, I think It would be marvelous if he were the bass fiddle. I could do a bass fiddle costume and then he could walk right out; that is if he's not too big. Can Joshua cast a 38 in the part? If the lighting was right, we could change the wood color to armour when he leaves the band. Jaynie, can you do that?"

"I'd have to see a sample first. It would be easier to make him a tree by projecting leaves on him."

"What are they eating at the banquet? You know catering a meal can be expensive."

"It can be something simple, John. Something you can put together yourself."

"This is a big show for me, Manny. I'd need two more men to do a dinner."

"Can't we hide the food somehow? How about doubling the bowls from the sleep walking scene? Then no one could see what's in them."

"Well, Harry, you see, Joshua and I sort of saw the banquet as a buffet. Everyone helping himself to food—all in tones of red—beets, carrots, jello, you know, a symbolic moment."

"I hope you let me do that scene in red, Pierre. Otherwise the stains from all that food will make the costumes a mess."

"Actually, Alvin, we saw it in white as a contrast to the red. But can't you use plastic that will wipe clean?"

"They never hang right."

"And the refection problem can be fierce."

"Let me ask Joshua. Maybe we can use red costumes and white food."

"Great. Potato salad lights beautifully."

"I remember once when Belasco served potato salad at a cast party and Fritzie Duncan, he was assistant director, threw this bowl all the way…"

"Manny, I could plant a great piece about potato salad, even get a recipe on the women's page with our cast eating."

"John, can you get us some potato salad in exchange for some program space? How about that deli down the block? They could use some theatre business."

"Wait, Manny, their potato salad is too red, too much paprika. We really need pure white potato salad, otherwise we lose the symbolism."

"I'll have to go with all white light. Other colors will spoil the purity of the potato salad."

"Hey, Pierre, I could do a potato salad hat and have the ghost rise out of the potato salad bowl."

"Say, if you have a big enough pot, I could put spots in there and when he comes out of the bowl it would be a really spooky effect for a ghost."

"...he threw it clear across the room and it hit Mable LaClare. She was Charlie's girl. Charlie looked at Mable all covered with potato salad..."

"I don't know if Joshua will be willing to make the ghost potato salad rather than a bass fiddle. It does change the concept."

"Before I leave, can you all get me a bio before Thursday?"

NEW COURT MASQUES

First published in **Lighting Dimensions**: October 1978.

At the American Theater Association convention in New Orleans this past August, members of one panel were asked to present papers on whether we were returning to a time of pure spectacle, such as the court masques represented, or whether we were moving toward a new stage-craft movement. After I had accepted an invitation to be on the panel a packet of information arrived including reprints of a three-part article I had written several years ago for Theatre Design and Technology. *I realized that these articles might be used as a background and target for the discussion but I could not use what I had said before. I would have to write something new.*

• • •

I think that we see only superficial resemblances in a comparison of today's theatre with court masques. But looking back, what a wonderful time it must have been then. The royal audience shared a mythology that made it possible in the theatre to say little but understand much. In our theatre we seem to say a great deal and understand nothing. And back then it must have been an energetic, lively, productive time. As Jacob Bronowski was fond of pointing out in books such as his *A Sense of the Future*, in twelve short years, when Galileo was at the height of his powers, the authorized version of the Bible, the first folio of Shakespeare, and the table of logarithms were published. One might even fantasize a return to that time, provided something could be done about plagues and enough ice provided for gin and tonics.

There was then, at least as far as we know, a tremendous faith in the powers of man to accomplish heroic feats of discovery, both geographic and intellectual. To me, the big difference between then and now is that in the midst of an unbelievable era of discovery, we have fallen into a depressive state that reflects a profound lack of confidence in man's ability to control his world.

But even more oppressive is that our present man-made technology appears so unmanageable that it becomes a threat to our existence. We seem to believe that with every forward step, we destroy more than we gain; we cause new diseases as we cure others; our new sources of power are our means of self-destruction; and as we look more deeply into the mind of man, we find no innate nobility, no common bond to make us brothers. Instead, we see only a biological link which is not strong enough to unite us, but allows us to retreat into an orgy of self-gratification, a gratification that has a bitter taste because it feeds upon itself.

When I wrote *The Scenographer as an Artist* about three years ago, I gave examples of many signs that to me indicated a shift toward a more conservative aesthetic, or at least toward an aesthetic which attempted to communicate. Since then, I believe we have been treading water in terms of artistic production. But as we rest awhile after the exertions of the modernist movement, there are indications, in what is now being said and written, that a program, an agenda is being developed in our society and in the arts. My perspective may be colored by my wishful thinking; I am aware of the danger. And I cannot give a time table for the arrival of new forms or messages. Nevertheless, there has been enough of a change in the past three years to provide an idea of what we might expect.

For one thing, artists, even those who work in private and non-communicative styles, are being told that they have chosen to make a statement. The late Harold Rosenberg put it this way in *Art on the Edge:*

> *Modern art is saturated with issues and ideologies that reflect the technical, political, social, and cultural revolutions of the past one hundred years. Regardless of the degree to which the individual artist is conscious of these issues, he in fact responds to them in choosing between aesthetic and technical alternatives...Thus, choices having to do with method in art become in practice, attitudes regarding the future of man. Hence, art in our time cannot escape having a political content and moral implications.*

If even an aesthetic of non-communication makes a statement, that message must have been a challenge to a past which demanded out of date and repressive conformity. But are we still fighting against that past? Suzannah Lessard, in a perceptive article in the *New Yorker* (July 10, 1978) speaks to this point as regards architecture:

Speaking generally, the sense of oppression by the past, of a deadly contest with a force that seeks to suffocate growth, seems to have faded away almost entirely...and to have been superseded by an urgent struggle of an entirely different sort—the struggle to control the future that we ourselves have created. If architecture can be used as a test case, that future form was greatly influenced by the conception of the past as antagonist. In a sense, we are now struggling with the stubborn grip not of the past but of the values created by the rebellion against the past.

This struggle is echoed in the other arts. In Stockholm in 1976, Saul Bellow, accepting his Nobel prize, said that writers do not "represent mankind adequately." Writers must "return from the peripheries." They must give the public "a broader, more flexible, fuller, more coherent, more comprehensive account of what we human beings are, who we are, and what this life is for."

Bellow is actually suggesting that there is some meaning to life. His words are a challenge to the self-doubt, nihilism and to the decades of a retreat into private forms of expression. One of the ways that life can have meaning is to believe that man has the power to order that life, to control his world. And it is precisely the belief that man can control his world that seems to be emerging.

The great microbiologist, Dr. Rene Dubos, recently said, "We must not ask where science and technology are taking us, but rather how we can manage science and technology so that they can help us get where we want to go."

This is, I think, the theme of the emerging change in our thinking. If we have the power to order our lives, then there are issues which must be faced. Questions of how to proceed will deal with the nature of man, the very thing that Bellow called for. Indeed, if we look around us there are signs that the debate has begun. Books and articles are appearing that at the height of the modernist movement would have seemed absurd. One example is the recent interest in ethics. Books like Sisella Bok's *Lying*, subtitled *Moral Choice in Public and Private Life*, are only possible where there is a faith in human choice, where there is an acceptance of man's ability to take responsibility for his actions, for his direction, for his future.

But a better example of ideas which are only now possible, only now taken seriously, is *On Moral Fiction* by John Gardner. Surprisingly, the critics listened to him with serious attention. He says this about morality in art: "Art is essentially and primarily moral—that is life giving—moral in its process of creation and moral in what it says." And morality to Gardner is "doing what is unselfish, helpful, kind, and noble-hearted...Moral action is action that affirms life. True art, by specific technical means now commonly forgotten, clarifies life, establishes models of human action, casts nets toward the future, carefully judges our right and wrong directions, celebrates and mourns." And he says of recent criticism "...it judges cynical or nihilistic writers as characteristic of the age, and therefore significant, and thus supports, even celebrates ideas no father would wittingly teach his children."

Whether or not we argree with him, a dialogue of alternatives has replaced a monologue of despair.

I would like to think that a dialogue will develop into a new humanism in our society. Others share that hope. Archibald MacLeish, in his book *Riders on the Earth* says:

> *There is no quarrel between the humanities and the sciences. There is only a need, common to them both, to put the idea of man back where it once stood, at the focus of our lives; to make the end of education the preparation of men to be men, and so to restore to mankind—and above all to this nation of mankind—a conception of humanity with which humanity can live.*
>
> *The frustration—and it is a real and debasing frustration—will not leave us until we believe in ourselves again, assume again the mastery of our lives, the management of our means.*

The question—one which I may appear to have ignored—is whether or not we will have a new stagecraft movement or return to court masques. My answer must be, court masques—no. A new stagecraft, YES! But that new stagecraft must be in the service of a theatre of ideas—and a moral theatre. We have so much to talk about. The real problems we face and the moral decisions we must make could occupy a legion of fine playwrights. And we must no longer be satisfied in a theatre that adds to the debasement of man by overwhelming him with technical effects, or which lulls him into further acceptance of his powerlessness by feeding only his senses rather than his mind.

Of course, scenography cannot communicate most of the messages that need to be said—at least not by itself. The role of the physical production in such a theatre is to add the emotional and sensual part that makes theatre an art rather than a philosophic treatise. A theatre of ideas is not a sermon. It is the expression of longings, fears, joys, triumphs, defeats, beginnings, and endings that make us human.

Humanity and morality are expressed in a partnership of all the arts of theatre. Therefore, scenography and technology must be partners not masters in a new theatre form.

ALL THE WORLD'S A SCREEN

First published in **Lighting Dimensions**: April 1979.

As with all frequently repeated experience, the effect is paradigmatic, effecting by analogy much beyond the immediately seen—indeed, all spheres of life where a fine and independent imagination matters. The much proclaimed ephemerality of television is no consolation; one might as well argue that since no one cigarette can in itself cause cancer, smoking holds no danger.

From *Daniel Martin* by John Fowles

● ● ●

Several years ago, while lunching with a colleague in New York, we discovered that we both came from Cincinnati. That in itself is not unusual. But it happened that we had both, at different times, lived in the same house and had even used the same bedroom. His family moved in after mine had just moved out.

The house was a huge gothic castle which had been built for a local bishop. By the time we lived there it had been broken up into several apartments. Even so, the dark woodwork, the high ceilings, and the oversized doors and windows made it impressive. My colleague had never forgotten this house of his childhood. Even now he would sometimes detour hundreds of miles to return to it and sit in his car looking at the house from the outside. (It is now a nursing home.)

The important thing in this tale is that since living there, no other home has satisfied him. His reaction is a good example of the effect, the lasting consequence of early experience.

Some people think that childhood influences are never overcome; that we are formed as children and when we express ourselves, we merely dredge up long-forgotten experiences. Certainly an artist's influences, surfacing in his mature work, can often be traced to his childhood. For all his years in France, Chagall was a Russian painter as was Picasso a Spanish painter.

My childhood in Cincinnati has imprinted on me a toned down gray palette. The solid stone of Germanic Cincinnati has constantly pulled me in one direction while my fantasy of being a romantic Latin has pulled the other way. No matter how much I may wish to be the product of Mediterranean skies, my earliest days are an anchor of solid and stolid weight.

One reason for mentioning the influences of childhood is that we do not often talk about why designers sometimes fail to achieve uniform excellence in their work. Obviously, we are more in tune with some productions than others. In some cases we are forced to create an environment which does not come naturally, which must be forced out with skill, perception, research, and experience. But even then, some jobs are more suited to our temperaments than are others. This is not to say that an outside observer cannot at times catch the spirit of a place or time more perceptively than can those who, through familiarity, have accepted the remarkable as common place. But most of the time we carry with us the baggage of early experience.

Not only do places influence us. The differences between generations, the time in which we grew up, leaves its mark. That is one reason it is difficult for teachers or union examiners to make sweeping statements about the creative abilities of a generation whose influences are vastly different from their own. Yet, if we are to guide and instruct, we must make value judgments based on what has formed us—and on an empathic view of current influences. Hopefully, there is a way to form a bridge between the standards of the past, made clear by hindsight, and the changing influences of a somewhat incomprehensible present.

I suspect that the single most pervasive influence today is television. I also suspect that we have as yet no real idea of how the TV experience will form future designers, artists, indeed everyone growing up with television as a constant presence. If in fact artists are forever gripped by their early experiences, what in heaven's name will be the effect on a child who spends the daily average of six hours watching television? These children are not absorbing the colors of fall foliage, the tones of a rainy summer day, the moon rising over snow banks. They are weaned on the colors and shapes of *The Gong Show*, the flashing numbers of *Sesame Street*, and the fake scenery behind *Wonder Woman*.

TV experiences are second hand. They are not from nature, not even from a great artist's interpretation of nature. They are from a designer who most likely has little control over what is finally shown on the screen.

Furthermore, the young viewer is at the mercy of the mechanics of a particular set and of the tuning tastes of his own family. There are probably millions of kids who think that skin is greenish and other millions who think skin is purple.

It doesn't stop there. What about a sense of scale? How does the size of a screen affect perception? And do the proportions of the screen establish in a young mind the ideals of shape that artists from Leonardo on have debated and searched for? What great mind determined that the TV screen should be as it is and not a half inch higher or wider?

I have lots of questions and few answers. But I cannot help comparing the visual arts' reaction to television to the reaction of other areas—language arts for example. Many teachers of reading have long debated the pros and cons of TV and a large group now feels that TV is a corrupting influence which must be countered. Perhaps the arts, lacking measuring tools such as test scores, are unable to perceive changes in the young. Or it may be that in the arts, the search for the novel, the acceptance of anything new, the recent belief that equates innovation with importance, has blinded us to the power of TV for doing both good and bad. Or maybe we have chosen to turn our backs on a problem we do not understand.

It seems to me that it is about time for some of us, maybe all of us in the visual arts, to become a bit wary of TV's effect on the future of our professions. It is important that, in John Fowles' words, "fine and independent imagination" not be less evident in the future than it is today.

THERE IS NO BLUE IN VLADIVOSTOK

First published in *Theatre Design & Technology*: Spring 1994.

When I arrived in Vladivostok to supervise the load-in of a set I had designed, I found the set unpainted. The major part of the set was a huge blue drop running from the theatre's apron to an upstage point where it curved upward and rose out of sight. The theatre management asked if I would mind if the drop were left as unbleached canvas. They explained that they had black, red, and green paint—but, in Vladivostok, no blue. Several days later, after calling other cities, they found some blue and painted the canvas.

ACT I: EXPOSITION

Vladivostok is in far eastern Russia, seven time zones from Moscow. At the southern tip of a short dangling tail affixed to the huge expanse of Russia, the city is surrounded by China to the west, North Korea to the south, and the Sea of Japan to the east. One million people live in Vladivostok, which, until 1992, was closed to foreigners and most Russians because it is the headquarters of the Russian Pacific fleet. San Diego, where I live, is also a Navy town on the Pacific; the two ports are sister cities.

In the winter of 1992, Efim Zvenyatski, producer-in-chief of the Maxim Gorki Dramatic Theatre, visited San Diego with a delegation from Vladivostok. He met Ralph Elias, artistic director of the Blackfriars Theatre. Even with a language barrier—Zvenyatski speaks no English and Elias no Russian—they hit it off and agreed to arrange an exchange. The first step would be an American production in Russia. Blackfriars would pay for expenses in the United States and transportation to Russia. The Gorki Theatre would cover expenses in Russia: set construction, housing and meals, and a small per diem.

After the main arrangements were made, a second theatre in Khabarovsk, 400 miles to the north, was added to the tour. (Most of what follows is based on experiences in Vladivostok because it was the primary stop.)

The Gorki and the Blackfriars are disparate. The Gorki has more than 200 full-time employees, 50 actors, a theatre that seats 1,000, and government support amounting to 95% of its budget. The Blackfriars, on the other hand, is a tiny Equity theatre with, depending on finances, three to five full-time staff, about ten actors and designers who regularly work with the theatre on a per-show basis, and no performing space, having been evicted from their storefront theatre in early 1993. Like most American theatres, only a small part of the Blackfriars' budget comes from government sources.

Elias chose to show Russia a play that had been one of his theatre's great successes a couple of years before: Beth Henley's *Abundance*. It has a small cast. It is very American. Set in the Wyoming Territory of the 1860s, the story corresponds to the Russian pioneer movement of the same time when Russians moved, not east to west but west to east, displacing the indigenous Asian peoples.

Having designed the Blackfriars' original production, I was asked to do the sets and lights for the Russian tour. In May, Elias, his board president, and I went to Russia to scout the theatre and make arrangements. We planned to return to Vladivostok with the company for performances in July and August.

The exhilaration of working in a completely new system and environment comes not only from the exotic experiences that are almost assured, but also from the contrasts that force us to reevaluate our own methods, techniques, and attitudes. The way the Russians treat artists is a striking example.

ACT II: "GET THE GUESTS"

I arrived before the rest of the company to supervise the set and lighting. The day the company flew in, I was invited to go to the airport to meet them. I said I would be glad to go if there was room for me. I did not know that several cars and a bus had been arranged to take a welcoming party of twenty-five actors from our host company. At the airport, three television crews were on hand to film the arrival. The Russian actors brought flowers for the Americans and the oldest actor and youngest actress had bread and salt for a welcoming ceremony. Seeing all this preparation, I told the theatre's director that the Americans would be overwhelmed by this greeting. He smiled and said," Yes, I know. It will help them give a better performance."

In our country, we give lip service to the appreciation and respect due our colleagues. We mouth pleasantries, clichés, and endearments that theatre people are especially fond of using. Wanting to be loved in our make-believe world, we pretend to love others. Actors, historically, have been treated shabbily. But they have a pay-off. Eventually they take charge of the stage and become the center of focus. Even that reward is now diminishing. New aesthetic principles are devaluing actors who now often support the physical production at great cost to their comfort or sense of worth. I have seen recent productions where actors were forced to duck through doorways, work in spaces so small they had to pass each other side-ways, walk like mountain goats on steep rakes, use steps that were more like ladders, or wear costumes that were more scenery than clothes. In the past, a certain percentage of actors—show girls or the second man in a horse costume—put up with being objects rather than subjects. Today, even lead actors become tools of the design. Noting this, the critic Mike Steele writes:

> Traveling the country, I feel some sense of loss as in theatre after theatre I encounter actors who don't have the feeling for language they once did, who don't project with the ease they once could, and who increasingly seem objects in some elaborate visual design that has pushed the Word into an increasingly subservient position.[1]

The plays I saw in Russia made me realize that their theatre is following our lead. Actors told me that they had fewer opportunities to play the classics and their productions were beginning to stress spectacle. And yet by a strange reversal, it was European design that influenced our own path toward the supremacy of visual over verbal art. The contradictions of international influences are a bit like a dog chasing its own tail.

Watching the Russians work gave me insight into how our theatre has been gradually changing, shifting its point of view little by little. It probably takes exposure to a completely different system to see how far we have moved over the past several years. In three issues of this journal in 1977 I wrote about the impending change from a theatre of words to a theatre of spectacle.[2] Those articles were controversial enough to be the basis for a panel discussion at the national convention. And that was even before Cats, Phantom of the Opera, Robert Wilson, or almost any production today that gets great public attention. What I had not foreseen was that, while we have been losing the poetic and intellectual use of words in our theatre, we have gained a wonderful freedom in our visual expression and we have, as well, given designers a more equal voice in the collaborative process. The trade-off has been more balanced than I had imagined. What bothers me, however, is that we have changed reflexively, not purposefully.

Look, for example, at the ubiquitous use of microphones, a change with few redeeming features. Russian actors still reach their audiences with the naked voice. The Gorki theatre is large with a proscenium more than forty feet wide and a depth of seventy-five feet. Actors use the whole playing area to great effect. True, those far upstage look miles away and sound miles away. But they can be heard whether they are speaking or singing. There are no microphones.

When I go to a musical here, I may as well be listening to a recording. Sometimes, from what I understand, the actors may even be lip-syncing. Like me, none of my theatre colleagues likes the sound produced by micing. Yet we no longer train our students to project because we have accepted the practice of sound reinforcement as inevitable. By making voices mechanical, I fear we have lost one of the most important qualities of live performance.

ACT III: "PUT OUT THE LIGHT"

Before my first visit to Russia, I had heard reports of how the Russians run a performance. The reports were true. Stage managers do not call cues in Russia. Instead, each crew member takes cues from the show itself, just like an actor. Stage hands and light board operators listen, watch the performance, and do their jobs at the right moment.

The Russians are able to do this because the crews attend many rehearsals so that by the time the show is ready to open, they have learned their "parts." Moreover, the lighting designer runs the light board. This system is possible because the actors, staff, and crew are employed full time, year round. I had no trouble accepting the reasonableness of operating this way until I considered that the theatre does fifteen shows in repertory, a different show each night. The crew must keep cues for fifteen shows straight. When I asked to see the paper work used for light cues, I was shown a sheet of paper with a few—very few—handwritten figures and reminders. I asked what happens when the board operator/designer is sick. The reply was, "I come to work no matter how I feel." But what happens if you are hit by a bus? "My assistant takes over...but it's not so good."

Their system works. During my first visit, I saw five shows and all were beautifully lit and well executed—and this in spite of equipment that was worn, broken, or non-existent. Many gel colors were not available, all the lamps were incandescent, and some lenses were cracked. The lights were a mix of all sorts, none of which looked at all familiar, plugged into a 200-circuit four-scene pre-set control board with about a third of the dimmers shot and several cross faders inoperable.

My job on that first trip was to figure out how to return with a light plot that could be used to hang, focus, and cue a show in a few short days. The Russians gave me a theatre ground plan but had no section, repertory plot, or instrument inventory. Thus, my challenge was to design a show with more than twenty scenes, work with incomplete information, communicate with a crew that did not understand my language (nor I theirs), and use a production system completely different from the one I knew.

My first method of communication, and ultimately most effective, was a series of photos of my model under different lights. I sent these ahead with my plans. Then I invented different charts which listed, in either words or numbers, the amount, direction, and color of light I wanted in each scene. Finally, I made a kind of magic sheet for each scene.

When I returned to Russia armed with all my brilliant charts and diagrams, the lighting crew was polite, even deferential, but I could tell they thought I was insane. They looked at the paper work, then pushed it aside. We talked and walked through the lighting design, discussing focusing and gelling. (I had brought a roll of various blues which were requested on my first visit.) Within a few days we had lit each scene and were ready for the cast to arrive. I had no idea what would happen with the cueing since there was not even an intercom between the stage and the booth.

The first tech rehearsal was more like a run through. The Russians had to hear the English words and watch what was happening. I paced downstairs in the house. The lighting designer and her assistant worked in the booth with the window open, listening to my instructions shouted up to them by the translator at my side. It was miraculous. More than half the time when I turned around to signal for the next cue, they were already going into it. They knew from the actions and tones of voice what should happen. They were artists. At the end of the rehearsal, we Americans faced the booth and cheered. The lighting designer said she appreciated our praise because she is taken for granted by her colleagues. We all understood that.

After several rehearsals, the lighting designer/board operator transferred the lighting information, cues, and timing, from odd scraps of paper to two full sheets. These certainly looked neater but seemed no less minimal by our standards. No matter, for the two women ran the lights as if they were organ players in a silent movie house; they didn't need much to help them play their board. The beauty of their control was that when an actor wandered slightly out of the light, (and I squirmed with annoyance), the women brought up a light just where it was needed. They were stupendous. I wanted to bring them home with me.

We talked about computer boards, which they had used on their company's tours of Japan. The women wanted no part of computerized control. They said that the kinds of shows they do and the number of lights they have do not require computers. I wondered how that may change if and when circumstances in Russia change and money is available for replacing old equipment. And I wondered what we have done to produce a system where some dolt who needs press only one button to execute a complete cue change often cannot get the timing right. What might result if we trained our technicians to think of themselves as artists?

Another incident spoke to the same difference between our systems. When the set was being installed on the stage of the Gorki, there was a question early in the process of where to put a piece of scenery. The technical director, crew chief, and stage hands looked to me, the designer, for an answer. I asked to see the ground plan. There was none on the stage but they sent for one. When that arrived I asked for a tape. There was none on the stage but they sent for one, apparently amused by my need to spot the piece precisely as it appeared on the drawings . They preferred to continue the artistic process by deciding how things looked at full scale on the stage, moving them around, and maintaining an aesthetic flexibility. Whereas they were ready to place scenery based on photos of the model and my feeling of the moment, I was relying on the instincts that brought me to make decisions weeks before in the studio.

My experiences with the lighting control and the set load-in, far from revealing the weaknesses in their system, showed me what we have lost on our path to technological proficiency. We have lost precious options by thoughtlessly accepting procedures and equipment that limit us as much as they free us. Yes, thoughtlessly—or as James Ogilvy says, "An occupational hazard among intellectuals is the stereotypical tendency to back absentmindedly into the future."[3]

To be very clear, I am not a Luddite arguing against technology. How could I do so when I am writing on a computer and will send the article to the editors on a diskette, having checked my references by modem and an international network. These are all wonderful machines, except when they break down, as my printer did last week, making me feel that I could not communicate. A handwritten letter? How primitive. I can imagine the editors' dismay if they were to receive this piece as a calligraphic manuscript, no matter how beautiful. We have grown dependent, the editors and I, on machines that speak to each other while our human contact diminishes.

Technology is providing us with tools which increase our powers, our reach, our efficiency. The artist Lynn Hershman refers to our tools as prosthetics, extensions of our bodies.[4] She lists telephones, automobiles, and computers as technology which allow us to conquer the limitations of our bodies. Yet it appears that the longer the prosthesis, the less personal the contact and the less real the perception. Besides distance, the amount of technology separating subject and object also makes a difference.

A computer expert tells me that people love electronic mail because they can say things in computerized messages that they would never say to another person face to face. Technological distancing works in our favor when we want to avoid personal contact. A more serious example is the use of smart bombs and other long-distance destructive weapons which supposedly take the guilt out of killing large numbers of people.

The very nature of theatre, however, is speaking to each other face to face. In a world of technological distancing, we crave the human contact we are quickly losing in the name of progress. Recently, in a series of medical tests, I carried a heart monitor for several days. At times I called an 800 number in New York and from California played back an encoded series of bleeps. A technician in New York used these sounds to produce a graph which was faxed to my doctor, completing a cross-country round-trip of startling technological prowess. However, I had become a number on the phone. No longer was a human hand touching me with a stethoscope or a human ear listening to my human heart. My doctor examined me in absentia, probably not even connecting my face with the paper before him. Is the impersonal graph more revealing and efficient? For certain information, yes. Do I miss the human contact? You bet. Is the examination incomplete without the doctor observing my skin, my eyes, the way I speak? I think so.

Looking at theatre history we see a succession of aesthetic swings to either distance the audience or to make greater contact. Each swing was prompted by the desire to use new technology or to overcome the distancing that technology had created. I was trained to use the proscenium stage which is able to produce the illusion of space. In the '60s I saw the growing use of thrust stages to foster audience contact. Now I see a return to the proscenium so that all the new technology needed to produce spectacle can be hidden from the audience and properly work its magic. We move toward and then away from audience contact, technology, illusion, spectacle, language, and reality.

In the larger society, interactive communication and virtual reality are today's hot topics. Interactive communication is a response to the cocooning effect, the isolation we feel at home with our TV and VCR. Can this technology ever be as interactive, as emotionally fulfilling as the best theatre experience? Isn't virtual reality actually a device for carrying us away from reality into a world of illusion? Perhaps we should call it virtual illusion.

Reality may have become too depressing or threatening in our society. However, when compared to the day-to-day reality of life in Russia and in much of the rest of the world, our existence is enviably pleasant. Yet it is here that I find we use technology, not only to minimize daily chores, but to distance ourselves from real experiences. I often see walkers or joggers on the beach with earphones on their heads. I wonder what they can be listening to that is more glorious than the sounds of breaking waves, sea bird calls, and the squeals of delighted children playing in the surf. While the option to listen to music almost any time and any place is wonderful, we must discriminate as to where, when, and also why.

In the theatre we must ask ourselves these same three questions about using new technology: where, when, and why? Balanced against any decision is the amount of technological distancing that results. Of course, trade-offs are inevitable. We might look for equipment that mitigates the distancing effect by permitting more human control, thus allowing for the vagaries of actors and crew. Real control is not pushing one button that sets in motion a complicated series of events that are then unchangeable no matter what else is happening. Control should mean running a show as an artist, the way those two wonderful women in Russia were able to do.

ACT IV: "THE FUTURE WILL COME OF ITSELF"

Like it or not, technology is changing our conduct and our way of thinking. Nowhere was that more obvious in Russia than in my host's attitudes about risk and fate. Here in Southern California, the reigning mythology tells me that if I eat a fat-free diet, exercise, and do not smoke, technology will keep me alive forever. If by accident I am injured or contact a disabling or fatal disease, I do not "...trouble deaf Heaven with my bootless cries, and look upon myself and curse my fate."[5] Instead I sue for malpractice or seek damages. Technology is perfect, human error is actionable. The Russians I met seemed to live for today. They took risks because tomorrow was uncertain and there was no available technology to make them imagine themselves immortal.

For a pipe smoker like me, Russia was paradise. I could smoke anywhere—in airplanes, in restaurants, and in the theatre—without raising eyebrows. Everyone else was also smoking, and drinking, and driving insanely. A good percentage of the cars in Vladivostok were late model used Japanese autos. (Unlike our tax laws, in Japan they make owning an old car prohibitive and the Russians buy not-very-used cars at a good price.) But the Japanese drive on the left and the Russians, like us, drive on the right. When I was a front seat passenger, inches away from speeding oncoming traffic, the term "death seat" took on new meaning. Seat belts? The cars have them but I never saw one used.

Perhaps the Russian acceptance of fate also explains a practice I noted with fascination and some horror when we performed Romulus Linney's one act play, *Akhmatova,* in the Gorki's 100-seat experimental Studio Theatre. This fourth floor room had two doors, one an entrance for the actors and one an entrance for the audience. At the beginning of the show, the house manager locked the audience door. Late comers could not get in and, no matter what, the audience could not get out.

Not surprisingly, given our histories, the Russian view of life, death, health, and safety are different from ours. They have one set of problems, we have another. I was interested in how their problems affect their writers. At a meeting of professors of literature, I asked the group to identify the major themes in current Russian writing. They told me that writers are still working the classic themes of mortality, morality, love, and honor. Our writers by contrast, seem to be tackling more specific problems, such as AIDS, feminism, and racial relations. The professors also mentioned that their biggest concern was that students come to them with a completely false view of history. Students in the United States come to college with no idea of history at all; they only know and care about the present. In both literature and history, the Russians are looking through the large end of the telescope while we are focusing on the details of life.

Might it be that the technology we enjoy, the relative ease of our existence, and the mass of information we command, have blinded us to the larger questions of life? Our artists seem to produce more and more work reflective of very personal concerns. Our tragedies stem not from some tragic flaw, measurable on a universal scale, but from the victimization of one or another sexual, ethnic, or gender group. We are looking into the wrong end of the telescope as we back into the future. The Russians still see the large picture.

The irony of our situation lies in the narrowness of our vision at a time when our sciences have created vast possibilities and frightening problems. In 1935, Charles G. Finney wrote *The Cir-*

cus of Dr. Lao, in which he described a man who had every conceivable operation to implant artificial parts. "One hundred years after he died they opened up his coffin. All they found were strings and wires."[6] Today, more miraculous implants are possible and some of us may be an assemblage of mechanical devices and parts from other humans. An article in the *New York Times* titled, "We Are What We Make," examines two new books which until recently would have been science fiction. The reviewer, Steven Levy, reports that according to the books being considered, "...the barrier between humans and machines has already evaporated."[7] The very titles of the books, *The Fourth Discontinuity: The Co-Evolution of Humans and Machines*,[8] and, *Metaman: The Merging of Human and Machines into a Global Superorganism,* are cultural sign posts.[9] Is our future evolution to be a merger with our machines? Will we then still be human? Do we have any control over our fate?

Beyond what may happen to our bodies in the future, we face a world in which computers may think like people, where genetic engineering, gene splicing, and cloning are possible, and where the information age has already made personal privacy almost impossible. Our technology is producing an ever-widening gap, not only between the rich and poor in our country but between rich and poor nations.

According to a scientist friend, as technology advances, the rules governing the ethics of his work change almost daily. For the rest of us, the speed of change causes us to blindly enter this not-so-brave new world with no philosophical, moral, or conceptual tools up to the task of charting our societal course. To create these, we must ask the question theatre has asked throughout the ages, "What does it mean to be human?"

To take control, I believe, two connected tasks confront us. First, to stay alive, theatre must rise above the themes of victimization to speak of the larger human condition. It must do so with metaphor, poetry, humor, love, and pity. It must question and comfort, inspire and console. It must break down the isolation of modern life and allow us, gathered together as a group, to face the uncertainties of who we are.

Next, as designers, technicians, and artists, we must control our tools, using them to create what is artistically right, not merely to show what is technically possible. As technology evolves, we must keep in mind the distancing that technology creates and resist the temptation to provide our audiences with only an experience. Large ideas and grand emotions are the stuff of theatre. We have a moral responsibility to harness technology in their service.

ENDNOTES

1   Mike Steele, "The Not So Empty Space," *American Theatre,* October 1993, p. 22.

2   Beeb Salzer, "The Scenographer as an Artist," *Theatre Design and Technology*, Spring, Summer Fall 1977.

3   James Ogilvy, "Three Scenarios for Higher Education: The California Case," *Thought and Action,* Fall 1993: 29.

4   Lyn Hershman, "Touch-Sensitivity and Other Forms of Subversion: Interactive Artwork," *Leonardo* 26, no. 6 (1993).

5   William Shakespeare, Sonnet 29.

6   Charles G. Finney, *The Circus of Dr. Lao* (New York: Viking Press, 1935).

7   Steven Levy, *New York Times Book Review,* 24 October 1993, p. 13.

8   Bruce Mazlish, *The Fourth Discontinuity* (New Haven: Yale University Press, 1993).

9   Gregory Stock, *Metaman* (New York: Simon and Schuster, 1993).

CAREER

[f. L. *carrus*]

1. A race course...

4. A course or progress through life...

● · · · · · · · · · · · · · · · · · · ·

Thoughts On Success And Failure While Moving

First published in **Lighting Dimensions**: September/October 1982.

"Son of a gun, that was the show where the leading man was fired at the dress rehearsal. I met my wife during this one. What was the director's name on this one? I remember the smell of his cigar but not his name. Oh! That portal, how they struggled to get it up. That little fat prop man was so good—when did he die? Look at that simple stool down left—it collapsed the first night out of town when what's-his-name sat on it. And those projection screens, they were fiberglass then. Every time they went up splinters of the stuff were all over the place."

I am looking over piles of renderings and sketches. Rolled plans, dusty on top and yellowing at the edges, add to the jumble and mess. I am moving. All the collected evidence of years of work lies waiting to be sorted.

My family has lived in this house for 14 years. My children know no other home. It is a funny Victorian house with so many angles and cubbies that there was always room to store those things we thought to be essential to life. Everything was saved because it was necessary. Throughout the house from the basement to the third floor studio, bits and pieces were hoarded away against future use.

Some day I'll need this little pipe, that chain, the bit of down spout, a sheet of glass, assorted screws and bolts. The list was almost endless. Of course, many of these squirreled-away treasures were actually used and just fit this or that project. Just as often, even though I knew it was somewhere, I was unable to find a needed piece and bought a duplicate instead.

In many ways the house and I were the same. Sometimes I could retrieve things in my head, but increasingly, of late, I couldn't find thoughts and ideas I had filed away. Moreover, as I began the necessary patching and beautifying of the house for our tenant, I felt that both the house and I were fighting a losing battle against decay. My identification with the house, and my accompanying sense of mortality, were not comforting. The house, already one hundred years old, will last far longer than I will.

One can really get spooked by contemplating the last gasp while pushing to complete a large task or meet a deadline, like moving day. That pile of renderings and plans didn't help. I found myself looking at it and thinking that none of it mattered to anyone but me, that one day even I would not be here to appreciate it. And worse, I did not like some of the work now. I could not imagine why I had saved it. It is a natural and important part of any creative process to step back and critically examine one's own work Add to the evaluation the criterion of the opinion of others after you have gone, and the result is chilling. So chilling, in fact, that if this morbid train of thought had been my constant companion I'd have walked away from the whole pile or set fire to the house.

There is, however, a bit of truth in those black thoughts, just as there is in the thoughts one has when the gods are smiling and every effort is a success. Some middle ground was needed for me to sort out what I wanted to keep and what got tossed. I wish I could report that a wonderful formula occurred to me, one that could help others evaluate their work. It didn't. My choices were purely subjective and idiosyncratic. I will admit that a small voice kept asking me, "Do you want to be remembered by this sketch or that rendering?"

A ton of conferences and articles have debated the way to preserve a record of theatrical productions in our times. Lighting design, especially, presents a problem for those who want to pass down a history of accomplishments. As I weeded out my own work I wondered again about the feasibility of the need for a complete record. How many of our productions will be of interest a hundred years from now?

Looking back to a hundred years ago, we are often fascinated by those works that we honor today but were considered failures by critics and audiences in their own time. Maybe today's theatre artists will suffer the same fate in years to come. If so, how do we possibly choose those representative works that are worthy of documentation? The artists themselves are no help. In interviews they constantly surprise by naming as their own favorite work one of their least successful, an effort more loved by the parent because it has been least loved by the world.

By heavens, it's a complicated problem. Try to outguess history and you get shafted every time. That sofa I sold to the used furniture dealer will be the exact piece we need in our new home. The rendering I throw out will be the one some yet unborn historian will need for his Ph.D. thesis. Tough luck Charlie!

It is even possible to take heart from daydreaming about the future. Fantasies crop to soothe the spirit. "Look out Ming Cho Lee, John Bury, Josef Svoboda, you might be riding high now—but in 2050, when my style is in vogue...."

But here I am still trying to make choices. What to keep and what to throw away? To do the job I have to be ruthless with myself. I choose a mixture of things. A few survivors represent shows that bring back particularly pleasant memories. Some designs are kept because the rendering technique pleases me. A few represent design problems well-solved or shows that were successful.

The chaff is thrown away and a strange thing happens. Instead of a sense of great loss there is the relief of being freed. Suddenly, the mistakes I've made, the bad designs, the poor solutions, have all disappeared. There is no record of my ineptness. Only my successes remain. I have been shriven of my artistic sins. The smoking pistols have been deep-sixed.

What if none of us carried the weight of our past mistakes? It's difficult to contemplate a life without guilt or failure. Once, hearing Frank Lloyd Wright speak I had the feeling that he at least never admitted a mistake. Gordon Craig may have had a similar ego. If I were mistake-free would I be a Wright or a Craig? Not likely.

I realize my mistakes are just as important to me as are my successes. (They better be, there are more of them.) I learned more from things that didn't go right than I did when all went as planned. There would have been no challenge if I'd known from the start that each new work would be great. Part of the excitement of any show has been pitting myself against all the variables that conspire to daunt me.

How refreshing it would be to see magazine articles that report on utter disasters. There could be profiles of designers that would chronicle lists of failures rather than the puff pieces we get now that only tell us of great achievements maybe all of us would learn something.

I am sitting here surrounded by what is left of my cleanup. The good things remain. Maybe I threw out the wrong pile.

Boo

First published in *Lighting Dimensions*: January 1979.

How long has it been since you stood at the end of a performance and booed the actors, director, playwright, or technical production? *That* long, really? But surely there have been times when a performance has been dreadful: the whole production was misconceived and painful to watch for anyone with standards. Well then, have you walked out in the middle of a show? Oh, you do that more often. Is that the extent of your willingness to show disapproval? But the actors may believe that has been some emergency which calls you from the theatre, whereas a few boos at the curtain call are, unequivocal messages of your contempt for their work. *What?* That would be impolite? You would never want to insult players who have worked so hard to put together a production?

I do not buy the argument that says it is not polite to boo at shows. The question of etiquette may become a factor as fewer and fewer people voice their unhappiness with a show, and booing does in fact, take on the stigma of a social no-no. But, like other forbidden acts (such as belching) booing was once a respected means of expression. If it has now become taboo, this is a change in social mores rather than a new awareness of some innate concern for the welfare of the human condition.

It seems to me that we have all become hyper-concerned for the sensibility of performers, not because of any regard for them, but because we are no longer able to tell whether or not they are doing a good job. The danger of this situation is that, like any commodity offered for sale, when there is no discriminating public, it is the cheapest, most easily arrived at product which is offered to consumers. When the public accepts low grade products, whether they be convenience foods or performances, the experience of something better is soon forgotten. Soon a new standard, a lower standard prevails.

In part, we must blame television for the harm done to public taste. The mesmerizing quality of the medium keeps our children rooted in front of sets, willing to watch anything that makes sounds and moves. There is no discrimination between commercials, cartoons, and *Masterpiece Theatre*. The set is never turned off in the middle of a program: the worst drivel is watched with the same absorption as the rare program of quality and substance. But the influence of television is only one part of the story.

Our whole culture has conspired to keep us from having discrimination. We are told time and again that the best new art of any kind always appears at first to be strange and unacceptable. If we wish to be philistine, and condemn the new and experimental, then we risk being considered road blocks to the juggernaut of advancing cultural history. Therefore, we tend to accept as genius any artist who had produced a work so outlandish that the critics and the media take notice. It is, of course, ridiculous to *dismiss* out of hand all new work; is it not equally ridiculous to *embrace* every new work that appears?

To have a developed critical sense, one must experience a succession of excellent performances. In most of our world, this is not possible. We do not get a chance to see a production of *Hamlet* every year, with a different actor playing Hamlet. As I think back, I have seen only a few *Hamlet*s, and I am in the business. How many does the average person see? But it is only through a comparison of one Hamlet to another that we can say that so-and-so is a great actor or what's-his-name stinks.

These days we read in the paper that a television series has discovered a super star who is one of the great actors of his day. Two years later we never hear his name. Put any of these media discoveries in a classic role before a knowledgeable public and then see whether they are great, super, or even actors.

There are times and places where booing is accepted behavior. One place is the opera in Italy, where a tenor or soprano puts life and limb in jeopardy by not living up to the standards set by discriminating audiences. But the opera-goer in Italy hears a *Rigoletto* or a *Butterfly* every year of his/her life! The aficionado boos, whistles, or stamps when disappointed but can be equally demonstrative when pleased. The Italian audience knows what it likes and is passionate about what it sees. And, in the Orient, an audience member at a classic presentation will walk to the stage and lecture an offending actor on the weakness of his performance.

These examples are both from societies which have maintained a classic form of entertainment. But we have in our country one form of performance at which the audience boos because they can measure the participants' proficiency: the sporting event. In this case, we do not seem to have the same regard for an athlete's feelings as we do for an actor's. A missed catch, a bad block, or a fumble calls down upon the player the wrath of spectators. There is no mistake about the public's acceptance or rejection of an athlete's performance.

We could assume that the sporting public is just ill-mannered, and that the audience for plays, ballet, and opera is refined and less willing to insult performers, were it not for the historical riots and battles instigated by theatrical performances. We can recall the Astor Place riot or more recently, the fuss at the opening of *The Plough and The Stars,* or many other passionate theatrical reactions. When the public is willing to take to the streets in defense of or in opposition to what has gone on in a theatre, we have an audience that *cares.*

The tragedy of our era, for those of us working in the theatre, is that we do not have a knowledgeable, passionate audience that cares. We are like master chefs catering to a clientele which can't distinguish between *pate de foie gras* and a Big Mac. Given a passive audience, we have only our own pride and self-discipline to push us to create at a high standard.

Sometimes even our co-workers seek the easy way out because of the public's indifference. This was the case, I believe, when I took some foreign visitors to see what had been a highly acclaimed production of *Porgy and Bess.* This was an American classic praised by the critics and I hoped that my friend, one of the finest designers in the world, and his wife, a distinguished actress, would be impressed by what we can do here in the States. The production turned out to be an embarrassment to all of us. The lighting was so horrible that I could not attempt to justify it even though I knew what had gone wrong. It was not the lighting design at fault. What had happened was that the electricians had simplified their cues by taking every dimmer to either full, half, or off. But even more disturbing was the follow spot which invariably cut off the heads of every singer. This is inexcusable, but there was not a murmur in the audience.

At intermission I asked to see the manager. He was not available. The lighting designer was not in town. I vowed to write or call him. *Mea culpa,* I did neither. In the rush of entertaining my guests, I forgot my ire. As I look back on this now, I see that it was as much a matter of my own laissez-faire attitude, nurtured by our culture, as it was the press of time. If I had called the designer, he would have rushed to correct what had been done to his show, had made him look bad and had denied other audiences a thrilling theatrical experience.

In any profession, including ours (if profession it is) there is competition, jealousy, artistic ego and insecurity. Human nature prevents us from being as candid with each other about our work as we could be. We still benefit from a few opportunities to exchange ideas and theories, but even these will disappear if we succumb to our society's acceptance of mediocrity. We must speak out and demand standards of excellence. We are, after all, the most knowledgeable and therefore appreciative audience for each other's work.

DAMN NEW YORK DESIGNERS!

First published in *Lighting Dimensions*: May 1980.

Not long ago I ran into Sam on the street of a mid-western city. I had not seen Sam for several years, and he looked older than he should have. He was graying, baggy of eye, and choleric of complexion. The double burden of hard times and compensatory drinking had soured his looks and his personality. The burnt out particles of his filament coated his soul and dimmed his light.

When I asked how thing were going he shook his head and cursed, "Damn New York designers!"

There was a bar nearby and after several doubles Sam was talking at a ten reading. His story was rather simple. He made a living teaching part time, renting out a small stock of equipment, and working the decreasing number of road shows that came in for a few weeks each year. But he had a degree as a designer and wanted to design. The shutters of his life were stuck in the closed position.

There were two local theatres, one ballet company, and one opera company. Sam had gone to each in search of design work. The answer in each case was the same, "Sorry, we only hire New York designers." Sam felt like a 110 strobe plugged into 220.

He knew some of the designers who came to town—he had gone to school with them and had done just as well as they had. He saw the work they did for the local companies and knew in his ferrite core that he could do as well as the farmed-in ringers. Often, they had little sense of the space in which they worked. They didn't know the local crews or the peculiarities of local practice. More importantly, they flew into town, spent as little time as possible, and flew out again, rushing to another pressing commitment.

Sam went back again and again to the managers of the local groups. To prove himself he offered to design a show for nothing. He would clean the stables or labor seven years. Even these classic methods would not move the managers. They needed a contact with New York.

There was powerful magic in New York. It was as if the travelers arrived from Manhattan carrying a piece of the true cross, a relic that gave professional legitimacy to a regional enterprise. The New York designer was not valuable for the work he did but as an amulet to ensure success. He became a mascot, a rabbit's foot, a lucky penny found in the gutter of Gotham. Sam could not compete with such powerful magic. His spells were candelabra lamps and theirs were carbon arcs.

I didn't have to ask Sam why he didn't move to New York. He had a family, friends, a house. He was a prefocus lamp, angled just the right way, only no one would turn him on. Besides, he hated big cities.

~

Several weeks later, the Union had a meeting to discuss new contract rates for regional theatres and opera companies. Harry, an old friend whom I hardly saw any more, was there speaking with great authority on the need for higher rates, better per diem, and first class accommodations. Harry was a 9 x 12 in the proceeding, focusing the group's attention on details that were clear to him. He had experience in just about every regional theatre there was. He gave examples to illustrate why certain contract provisions were necessary and some were not.

After a vote was taken, I caught Harry out in the hall. "How about a drink sometime or lunch?" Harry said he would have to look at his date book.

"Can't then…be in Indiana. No, not then, off to Florida. Vermont the next week and then over to Ohio."

It became clear that he spent four-fifths of his time away from New York. I asked him how his family liked him being away so much of the time.

Harry told me that he and his wife had experienced some rough times, but that they had come to an agreement. She and the kids had moved to Connecticut where her family lived and where she would not feel so lonely. Harry spent as much time there as he could.

"Then you are seldom in New York?"

"Not much. I have a studio in the 30s…more for mail and answering service. That New York address is important. When I'm not on the road I try to stay in Connecticut."

"Do you see many shows here?"

"3 or 4 a year, ones I have to see because someone gives me tickets. There are always things I want to see, but never have the time to get to. I am going all the time. This kind of life is a killer. I'm a human tracer. I wish there were some way to stay put."

Harry and I never did have lunch. He was always too busy. But every time I read a review from some far off theatre and see Harry's name in the credits I think of Sam and the way he said, "Damn New York designers! " Maybe there is no mystery to life—it's just a sine wave.

E.G.O.

First published in *Lighting Dimensions*: February 1979.

One learns a lot from reading *Lighting Dimensions*. For one thing, over the months, it seems increasingly clear that feelings run high among lighting designers when one kind of design, theatre, film, disco, circus, etc. is said to be better, more distinguished, more lucrative, or more creative than other kinds of design.

It was with great interest therefore, that I received information from E.G.O. (Evaluational Grouping Organization) which attempts to settle for all time these inter-professional squabbles. It is important that designers to be able to locate themselves in the pecking order of professional life, according to E.G.O. Having a secure standing and a realization of achievement confirms in the designer a sense of the self. In other words, "I grovel, therefore I am."

To arrive at a clear understanding of each designer's position requires all the resources of the social sciences. There is, in each case, considerable material to be evaluated. But with conscientious effort it is now possible to formulate tables of relative standing based on mathematical equivalents. Some of the standards which appear in this article may seem subjective, but that is only because there is not enough room in this small space to publish a complete list of the numerical ratings. A complete and comprehensive list of variables will soon be available in a twelve volume set from E.G.O.

ASSOCIATIONAL POINTS

A numerical equivalent is assigned to the star of each production. The lighting designer computes his associational score by using the table of names and dividing by the factor given to each of the following circumstances:

• Does the star know the designer's first name? Factor of 1.
• Does the star know only the designer's last name? Factor of 2.
• Does the star purposefully not know the designer's name? Factor of 5.

Example: Lawrence Olivier = 93. He only knows the designer's last name, divide by 2. Total associational points = 46.5.

Example: Rock Group, *Oedipus and the Mothers* = 3. Designer is known to group by first name, nick name, and several affectionate diminutives, divide by 1. Total associational points = 3.

Recognitional Points

If designer is known to the average man-on-the-street, award 5000 points. If designer is known to 30 other designers, award 25 points. A guest shot on a TV talk show is worth 1000 points but is only good for 45 minutes after the end of the show. A profile in the *New Yorker* is worth 75. Please note that profiles in the *USITT Journal* or *Theatre Crafts* are given no points because the *USITT Journal* only prints articles about Josef Svoboda and *Theatre Crafts* only publishes accounts of fictional German designers with hyphenated names.

Review Points

On the face of it, reviews are easily given a numerical value depending on where they are printed. A review in the *New York Times,* the *Washington Post,* or the *Tombstone Epitaph* is worth 100 points. A review in the *National Enquirer, Playboy,* or *Back Stage* is worth 200 points.

But there are other factors involved. If the designer's name is mentioned once, it does not matter if the review is favorable or not. If the designer's name is mentioned two or more times, points are assigned on a geometric progression. Points are subtracted on a scale of arithmetical progression for unfavorable reviews but this subtraction is made from points accumulated by multiple mentions.

The place in the review where the designer's name is listed is important. Any time a designer is mentioned before the star or director, subtract double the accumulated review points. If the name appears in the first quarter of the review, multiply by 3, if in the first half multiply by 2.

Finally, points are assigned to the adjectives used to describe the designer's work. Examples: glorious = 3, pristine = 4, neat = 1, murky = −5, coprolitic = + to −5, depending on the show.

Control Points

It is easy to compute control points because no table of equivalents is necessary. On the plus side, the designer simply adds up the number of human beings he can give orders to. Added to this is the number of people the designer can fire or have fired. This number is multiplied by 3 and added to the total.

On the minus side is a list of those whose orders must be followed by the designer. Again, a list of those who can fire the designer is multiplied by 3 and subtracted from total points accumulated.

LISTING POINTS

The placement and size of the designer's name in programs, posters, and display ads is assigned points as follows: before or larger than any other name, 1000 points, same size and among other names, 200 points, smaller and last, 10 points.

Almost always associated with educators, academic or organizational titles merit certain consideration. Examples: instructor = 0, Asst. Prof. = 1, Assoc. Prof. = 1.25, Prof. = 6, president of professional society = 10, Contributing Editor = 500.

PENNY POINTS

Penny points are merely a correlation between the designer's place in the pecking order and his tax bracket. Points are awarded on the basis of the final figures on *Form 1040*. (If these are not honestly derived from actual income, the designer sacrifices a certain amount of professional standing.) For every percentage point of taxes, 1000 points are given. Example: 30% = 30,000 points, 50% = 50,000 points. Income derived from other than professional activity is figured into the total because it demonstrates a wisdom that deserves to be rewarded.

It is obvious to every right-thinking person in the lighting field that the E.G.O. system is a long overdue and a valuable tool for knowing where one stands. Further advantages will accrue when *Lighting Dimensions* publishes a directory of all names and numbers. At that time, countless hours will be saved for hostesses in both invitation choices and seating arrangements. Equipment manufacturers will know which inquiries to answer and which phone calls to accept. In short, it will bring order to an otherwise chaotic situation.

I WORRY

First published in *Lighting Dimensions*: July/August 1984.

Laboro ergo sum. I worry therefore I am. I worry that having barely passed Latin, my attempt to use the same form as *cogito ergo sum* will get me in trouble with scholars who will then demonstrate to the world how stupid I really am. Not only will no one read my articles, I will be the target of public ridicule. People will laugh when I say profound things, knowing that they are either not profound or that they are stolen from someone else.

In short, I worry about being found out, about being revealed as a small boy in an old man's body, trying to play at being a professor, an artist, a businessman, even a father. I worry about how hard it is to be grown up.

I worry therefore I am is very true. I am not speaking of fear. That is a separate problem, I am not speaking of stress, that new buzz word that has everyone from tight rope walkers to monks measuring their own bodily rhythms. Fear and stress are mostly the result of real problems and reactions to actual events. When a 64-wheel semi carrying explosives jumps the highway divider and comes straight at me I do not worry. When I have drawings for two shows due in six hours and have not started either show I do not worry. Instead I am gripped by fear and stress, which produce real bodily changes: the adrenaline flows, the heart beats faster, breathing is increased, perspiration starts, pants are wet.

Worry, on the other hand, is usually produced by imagined circumstances which other people cannot begin to understand. What if, I worry, on the way to a meeting with the director, a car going in the other direction drives through a puddle and splashes muddy water through my window and ruins all my drawings. The director, knowing nothing of my thinking, wonders, "Why the devil is that guy spending fifteen minutes taking his drawings out of twenty-six watertight plastic bags?" Private worries explain strange behavior by otherwise normal people. How else can we account for the light board operator who never works unless he is wearing rubber wading boots, or the actress who will not go on stage without concealing a table of international long distance telephone rates inside her bra.

I do not want to imply that all worries are trivial or silly. Some worries are important enough to form the basis for great philosophical theories. For example, we can worry about the meaning of life without suffering fear or stress. We can even classify people by the quality of their worries. Since we all worry, we all fall somewhere within a spectrum of worriers, from profound to trivial. The measure of a man is how much he knows; how much he owns; how kind he is to family, pets, and inferiors; how healthy and fit he is; how tall; how sexually attractive; and how profound his worries (not necessarily in that order). The measure of a woman is exactly the same except for height So far, there has been no demand of equal height for equal work.

Classifying a person according to the quality of his worries is difficult because we think we should not talk about our worries. Admit it. If someone asks how you are, even if you have spent three days worrying, you never say that you are worried. You might say you have a cold, have just discovered you need root canal work, that your wife has left you for a hang glider test pilot, but never, never, would you say you are worried about whether or not Liechtenstein has the bomb. Can you imagine this exchange:

"Hey Joe, haven't seen you in ages. How are things?"

"Pretty good. But I'm worrying about the theatre not having any great new playwrights."

Poor Joe won't be asked to have lunch until rabbits have horns.

Worrying is not considered fun by our society—except by mothers, the one group that is allowed, expected, encouraged to worry. For the rest of us, we do anything to keep from worrying. Much of the work we do is merely filling time so that we do not worry. We play tennis, do aerobics, collect stamps, or attend garage sales—just to avoid a good worry. Worrying is not respected as a recreational activity.

I'll tell you what really worries me. Because we are not supposed to worry, our president is running a reelection campaign based on the premise that everything in the country and world is so great we do not need to worry. He is saying there are no real issues, therefore, we the electorate need not worry. We are supposed to look at him, see that he is not worried, and indeed not a worrier, and then happily go back to our frisbee contests. It seems to me that if he is not worried he doesn't know what's going on. I worry because he doesn't worry.

I worry that we are not willing to accept our share of the world's worries. I worry that the Japanese or Russians are taking over our share of the world's worries. In the future battles for commercial and philosophical dominance, it will serve us ill to be champion paddle ball players while others have become expert worriers.

Two kinds of pressures are at work on our worry systems. On the one side our government and society encourage us to keep busy doing almost anything rather than worry about important issues which make us human, such as hunger or nuclear genocide. Our government makes us less human, less concerned, less feeling. At the same time commerce forces us to worry about unimportant things: bad breath, split ends, and seeing our reflection in the dishes we've just washed. The message is that worry is bad, so give us your money and worry no longer.

Each stage in life has different worries. Can you classify in chronological order the following examples? Is that a pimple starting under my chin? Can I get rid of it before Friday night? Will that new suntan base cover it up?" "Are interest rates going up? Should I sell GE and buy gold?" "If I move to Florida, will the kids ever visit? Who will take me to the hospital if I get sick?" "If I close my eyes, will bears come out of the closet and eat me up?"

Please note that many worries involve questions and have causal clauses. If such and such happens, what will result? Some of the triggering events may be real and likely, others are impossible and imaginative. The more imaginative are contained in the worries of children and artists who are therefore able to worry about more wonderful things.

Worry abhors a vacuum. It will fill any time not used for other purposes. Try to sit quietly without reading or watching television and not worrying. It is impossible. For most people, the value of reading and watching television is that it blocks out worrying. If by chance you start to worry while you are watching television, either the program you are watching is terribly dull or your worry is terribly strong. (If the program is causing the worry, it must be an educational channel.)

Theatre people have special worries which vary depending on their jobs. We can set up a worry formula according to how much physical control of a production the worrier has. Theatre people who actually do things, push buttons, move scenery, zip zippers, have few worries and less time to worry, since their time is filled with real activity. However, when one depends on others to get things done, the open time for worry expands. Producers and directors, consequently, have the most time to worry, although producers retreat from good honest worrying by having lunches, talking on the phone, and reading columns of numbers. Directors have no physical activity to protect them against worry except exercise on the casting couch, and that causes other worries.

I worry a lot. Sometimes I worry that I worry too much, but most of the time I sort of enjoy worrying. Wearing my hat as a professor, I am paid to worry. The saying should be "worry or perish" because worrying is a sign of wanting to find answers, exactly what a professor is supposed to do. Professors are said to be absent-minded. That is not true. They are only worrying. A typical professor's worry would be about the definition of "scenography."

I have been worrying recently because I saw some wonderful productions at the Olympic Arts Festival in Los Angeles: the Pina Bausch Wuppertaller Tanztheater, the Piccolo Teatro di Milano, and the very best, Le Theatre du Soleil. Two of these have lighting effects so simple they could be accomplished on a couple of portable six packs. That fact is enough to furnish months of good worrying. The third, Piccolo Teatro did a whole production of *The Tempest* without one bit of front light. There was side light, down light, and back light, but not one face was ever lit from the front. I left the theatre having never seen a face, only silhouettes. Every principle of lighting practice was shattered. Now I must worry about rules.

As I get older I find there is too much to worry about and too little time to worry. I am considering hiring a worry assistant, some one akin to a research assistant. I worry though, that it would be difficult to check on the quantity and quality of his or her work. Maybe we must do some things ourselves.

Now I must begin worrying about what to write about for the next article.

IN 25 YEARS

First published in **Lighting Dimensions**: December 1979.

When I was growing up in Cincinnati, my great aunt, more than other members of the family, shared certain childish enthusiasms and took me to parades and circuses. She died a few years ago at the age of 101 and thus severed an extraordinary link with the past. Her mother had been taken to hear Lincoln speak in Cincinnati, and she in turn took my mother in 1908 to see Taft receive notification of his election to the presidency. But more than anything, she was for me the embodiment of an age that saw almost incomprehensible changes in technology.

She was fond of reciting a poem she had learned in school about a young man, Darius Green, and his flying machine, a comic tale which made fun of the idea that man would ever fly. Yet in her life she saw the advent of airplanes (in her old age she rode on one to attend my wedding) and lived to see men on the moon. Moreover, she saw the introduction of electric lights, telephones, radio, television, computers, and automobiles. Her tales about life without these machines fascinated me as a child and as an adult.

Carl Sagan, in his book, *Broca's Brain,* sets up a comparison of time and movement that emphasizes the speed by which the world is changing. For millions of years men walked at a few miles an hour. Only several thousand years ago, horses were domesticated, and man was able to travel at 10 or 20 miles per hour. But in our own century alone, "...products of human inventive genius have enabled us to travel on land and on the surface of the waters a hundred times faster than we can walk, in the air a thousand times faster, and in space more than ten thousand times faster." My aunt was alive when the Pioneer rocket was launched. It is traveling in space at 43,000 miles per hour.

The *New York Times* carried an article a short time ago about the "futurist," F. M. Esfandiary. He looks ahead to a world powered by nuclear fusion, geothermal, and hydrogen fuels that will create a prosperity now undreamed of. But his most startling predictions are for the science of medicine. "Anyone alive in 20 years will be alive in 200 years, and if you're alive in 200 years, you'll be around forever." Given the lesson of Darius Green and his flying machine, we scoff at such predictions with less confidence.

One other startling figure shows that we will all be affected in dramatic ways by the speed of technological progress. Our scientific knowledge is now doubling every eight years.

LIGHTING IN THE PAST

All of this caused me to look back over the changes I have seen in theatrical lighting. *Theatrical Lighting Practice* by Joel Rubin and Lee Watson was published 25 years ago. I got out my copy, bought as a student, and looked at what was then the state of the art. Resistance dimmers still held sway even though the first electronic preset boards had been introduced. Television studios were using auto-transformer controls with mechanical interlocking. Incandescent and arc lights were the only available source, and gels were still gel. No one dreamed that in a few short years there would be computer memory systems, miniaturization, or the variety of light sources we have today.

There are still places that use resistance dimmers or auto transformers. Therefore, one of the minor debates among academics concerns whether or not to train students with computers and then send them out into a world of tin can lights and bread toaster controls. Looking at the changes that have come so fast, the proper question might be how to train students to keep ahead of the developments that are likely to come in the future. Some of us, trained not so long ago, have trouble understanding what goes on under the hood of the machines we use today. With an increasing speed of technological possibilities, what we teach now may be obsolete before a student leaves school.

A FEW PREDICTIONS

I see no way of teaching things we cannot know about, but it might be fun to try to look into the future and predict what may be the state of the art in another twenty-five years. We can play "futurist" by looking at those fields from which lighting borrows its technology—computers, electronics of all sorts, and pure physics. Then, we can imagine what would be valuable for use in the theatre. Add a dose of pipe dreams, stir well, and we end up with some predictions of questionable reliability—but then, who knows?

Looking at light sources first, it has always been known that a point source would be of great advantage. Lamps are getting smaller and may get very small, but they might also change in other ways.

Maybe by chemical or electronic means, lamps may become color variable. Imagine dialing the color temperature of a lamp, putting it into the computer control, then changing the color of each lamp during a show. It is already possible to mechanically change direction, focus, intensity, and color, but mechanical color changing is limited to the gels in a magazine. Covering the entire spectrum by a new means and combining this with the other mechanical and electronic process might let us light a show with a dozen instruments.

The new lamps would, of course, be very efficient. What little heat they did produce would immediately enter the theatre's HVAC system.

LIGHT PRODUCTION

Thinking like this supposes the continued production of light by methods we have known. Maybe we are in store for a jump to a brand new way of making light, the way electric lights were a jump from gas. In basic research these days, scientists are "peeling the skin from the onion" and finding that each time they think they know how an atom is put together, a new layer of smaller particles turns up. We already have quarks and charms of different colors, strong forces and weak forces, gluons and anti-matter, besides the neutrons and electrons that we learned about in school. There is even a renewed interest in what was once a fanciful idea that our whole universe is merely one atom in an incredibly larger object. Our universe may be part of a vast table leg while within every atom in our world is a tiny universe containing stars and planets and creatures so small they are hard to imagine. What is important to us is that as man discovers more and more about the atom, it might be possible to produce photons in a new way.

A particle gun might activate the atoms of air around a performer, causing a glow around the person. We would be creating light at the object rather than sending light to it. The performer might even carry his own light producing apparatus.

We could have projection systems that work not on flat screens but on curtains of magnetic fields through which actors could move. No more projecting on smoke—just images existing in air. Using the principals of holography, we should be able to have three-dimensional images, very solid looking, any place we want them. There wouldn't be much need for scenery unless, for example, a door had to be slammed.

Controls for these new systems might be automatic. No humans need touch the dials. The problem of following a performance by live actors whose timing changes each time around could be solved by having voice activated controls. The control board would learn the play and receive commands from actors saying their lines. Or it may be that sensors on stage will give signals to the board so that light will fill an area as soon as an actor enters it. With all the new devices, an actor need never again be in a dark hole (we speak professionally not personally).

New controls may send signals by light rather than electric current and instruments, powered at their own position, and responding to commands by radio. Cables and wires would be unnecessary.

Hopefully, the industry can arrive at some standardization which will allow a show's cues and a total set-up to be sent from one theatre to another. To do this, the computer in one theatre will talk over the phone to a computer in another theatre. Using more complex computers, other tasks will be assigned to them. Management will handle payrolls, tickets, inventories, billing, and tax forms with the same computer that runs the lights. One hopes that errors in the system would not result in the IRS getting cue sheets for *Hair* instead of W2 forms.

SOLAR LIGHT

The control of solar energy may bring ways of storing light. It seems silly to convert light to electricity and then later convert it back to light. Therefore, solar light batteries will fill the basements of theatres and captured light will bounce back and forth between perfect mirrors until that time when it is released bit by bit to light a show. If this system becomes difficult or if energy problems become more acute, we might have to go to direct solar energy. The Greeks used to do that for all their shows.

All of this speculation has made me giddy, and I must return to reality and try to figure out how to keep the lens in that one fresnel from shattering when the soprano hits her high note. But there must be readers who have their own visions of what the future holds, who will pick up the baton and let *Lighting Dimensions* know what they foresee in 25 years. The publisher may even offer a prize for the most accurate prediction—to be awarded a quarter century from now.

MURDER, HE WROTE

First published in **Lighting Dimensions**: January/February 1986.

My friend Simpson looked like a bleached car mechanic in his prison fatigues. His face was gray. His clothes were gray. The wire mesh of the visitor's screen drained more color from the man who was formerly flushed of face and brilliant of attire. Not many lighting designers are put in jail (although set designers and costume designers often wish on first stars to have them there). Moreover, there had never been a case of a lighting designer charged with murder. But here was Simpson, lighting designer extraordinaire, known for his patience and stability, talking to me three days before the state would take his life.

"I'm not a bit sorry. I know the judge would have given me life if I had been sorry—but I'm not. Not even now. I can't tell you how much I had to do it. It was a passion that no one can fully describe in plays or poems. And when it was done, there was a feeling of exaltation that has made all the subsequent pain worth it. I know that sitting in the chair I will still believe that given the chance again, I would do exactly as I have done.

"For several years I didn't know who he was. It was strange. I used to read his reviews of my shows and, inevitably, there would be a line or two about my lighting. Every mention was favorable. No. I should say they were raves! Why be modest now? 'The lighting by Simpson was just right.' 'The lighting was spectacular and added immensely to the production.' 'Simpson's work adds to a production each time. Lucky is the show that lists him in the credits.' You see, I remember them all.

"I couldn't help but appreciate all the good reviews. I had never met the man. In fact, I didn't even write him a thank you note after several of the raves had appeared. After all, he was pleased with my work and a personal contact might just spoil it all. I even began to think his reviews were simply my due for good work. Why should I thank him for what was only the truth, for what I deserved?

"The good notices continued. My personal good luck critic was watching over me like a guardian angel. He was certainly a man of taste and good sense. If others in the shows I worked on had terrible things to say about him, I ignored them. They were no doubt bitter about some well-deserved jab. My favorite critic was fair.

"Then one opening night it happened. Standing at the back of the theatre I noticed the director and producer arguing. The producer was physically holding the director's arms, calming him. Finally the director was pulled out into the lobby. I followed. The director screamed, 'That son-of-a-bitch, I'll wake him up. I'll show the bastard he can't fall asleep at my show.' The producer soothed him. The director's wife came out to the lobby trying to quiet her husband, who could be heard in the theatre. She took him to the bar next door.

"When I asked what was going on, the producer told me that my critic, my angel, was fast asleep in the sixth row aisle seat. The producer was not at all surprised. 'It will be okay,' he told me. 'I've left the bottle of scotch under the seat, the way I always do.'

"He was right. The next morning a stupendous review appeared in the paper. I got my usual rave.

"You can imagine what it did to me. All of the praise, all those good notices not only meant nothing, they turned to negatives in my mind. Then I had my first major bout of self doubt. It took me weeks to come out of it. In the process I screwed up everything I touched.

"After a while I was back to normal, because I told myself that one night must have been an aberration. Why was I torturing myself because of one slip? The man must have been overworked, or tired from a red-eye flight. Who knows how many real reasons there could be to excuse him that night?

"But, of course, it happened again. It was the night Nelly Merwany got sick. Replaced by an understudy on the opening night, Nelly still got a glowing review the next morning. From then on I noticed all sorts of things I never wanted to see before. He must have been writing his pieces before he came to the theatre. And no one seemed to care.

"Then there was the case of the New Star Playhouse. They were getting great reviews from him and I went to see some of their shows. You cannot imagine how terrible those shows were. A short time later it was announced that he was to be the translator for one of their upcoming productions.

"The list goes on and on: talk of girls and drinking, free trips, hotel rooms, drugs. All of it provided in exchange for a good word from him. Maybe it wasn't completely true. You know how people exaggerate such things. I had stopped paying attention to what the critics said and told myself that the steady stream of good reviews from him was no more important than a lot of bad reviews would be. Yes, the good ones kept coming. It was as if he had my name in a computer that was programmed to mention me only in positive ways. Maybe he really did have a computer. Who knows?

"One night, as was sure to happen, I met him. It was at an opening night party for Berny Sud's new play. Jim Klaner and I were talking when he walked by. Klaner grabbed his arm and introduced us before I could stop him. I tried to be diplomatic. I said how much I had enjoyed his writing and how grateful I was for all the times he had mentioned me so favorably. He looked at me with the blankest stare. I saw that he had no idea who I was. For years he had written about me and now he could not remember ever hearing my name.

"I don't know why, but right then I asked him if he didn't have a review to write for the paper next morning. 'Oh,' he said, 'I've phoned it in already.' Here it was, less than thirty minutes after the curtain had come down and he had written his article, or so he said. So much for thoughtful evaluation.

"The next day I ran into Berny Sud. He told me how surprised he had been on that opening night to find that my critic and I did now know each other. He had assumed that I was on the man's pay list. He said everyone assumed it.

"That was the last straw. He had to give me a bad review. I wrote to him asking for an appointment. He never answered (but the police later found my letter).

"There was only one thing to do."

Simpson died as a true man of the theatre. As they strapped him in the chair he looked over at the control booth. His last words were, "Let's read that circuit up at ten."

LIGHTING THE FUTURE

First published in *Lighting Dimensions*: July 1980.

Architects, like some lower orders of insects, are able to adapt to great environmental changes and survive throughout long fallow periods of deprivation. They switch from one form of work, say residential, to others, possibly governmental, depending on the economic climate. Or, switching metaphors, they are like migrant workers, harvesting whatever crop is in season.

The lighting industry, grown during the disco era to unimagined proportions, must look to a disco-less future in which flexibility and imagination will permit the same kind of diversification that architects practice. Lighting professionals must now learn to adapt.

DIPLOMATIC LIGHTING

One area that is completely unexplored is diplomatic lighting. (This is not lighting designed for international conferences or meetings. While lighting could certainly be devised to make meetings more congenial, put representatives at ease, and smooth the way to agreement, I predict a more direct approach to the attainment of national policy through lighting.)

LASERGRAMS, $30,000 PER WORD

Since theatrical lighting has always been in the service of fooling an audience, making it believe in the unreal, it is not a major jump to make a whole nation believe a fiction. (Some cynics would say this is the main job of politics.) We already have the tools by which psychology, combined with lighting, can manipulate opinion.

Imagine a laser programmed to write a message. The rapidly moving beam appears to be a continuous line of light. Tinkerbell in the current Broadway production of *Peter Pan* is just such a laser beam. Besides flitting from place to place, Tinkerbell writes like a huge light pen on the scenery.

Imagine, now, an undeveloped country populated by religious, superstitious, and scientifically innocent people. In the cloudy night sky a message in light appears, moving across the sky. More than a mere projection, this is brighter by far, a message written by a heavenly hand. "God says release the hostages " "Drive the Russians Out." "Castro is a Devil." "Carter is the Messenger of God."

LIGHTING FOR SEX

Until now, lighting effects for sexual acts have been primarily mood inducing. The psychological state required for utmost fulfillment varies greatly and can be enhanced by many variables, not the least of which is lighting. Extremely shy people may prefer complete darkness, flickering light may be the choice of drive-in movie enthusiasts and nature lovers may choose the brightness of sunlight. While all of these lighting conditions are selected for psychological reasons, there may also be a correlation between light and the physiology of sex.

It is well known that light affects the body in many ways. It can cure certain liver conditions in premature babies, cause variations in female maturation and menstruation, and even bring about skin cancer. Lack of light affects miners and others working in darkness. Would it not be reasonable to assume that light in the proper intensity and wave length could create physiological conditions favorable to sexual activity?

With proper research, lighting devices could be sold as sexual aids. Imagine the available market for lighting kits which would include: lights, manuals, ceiling mirrors, and assorted paraphernalia. What a boom they would bring to the lighting business. New empires would be erected, a fitting climax to operations that have often been in the hole.

This project should be started at once. The research alone makes the doing worth while.

ENVIRONMENTAL LIGHTING

I recognized another possibility for lighting industry expansion when I visited a home in which a large glass wall looked out on a lake, forests and a small island. As the sun went down and the view became darker, certain parts of the scene stayed light. The effect was surreal. When the sun was completely gone I realized that artificial lighting (if we in the business should use that term) was brightening parts of the island and forest.

There is a certain magic in seeing large buildings, impressive monuments, bridges, and ocean liners lit at night, The scale of the objects, the number of lights, and the lighting skill required bring "ohs" and "ahs" from appreciative viewers. But, as common as it is to light large scale man-made objects at night, natural forms have only been lit on a small scale. Just think of the market for large outdoor lighting projects.

I, for one, would be interested in a contract to light the Rocky Mountains. Likewise, the Grand Canyon is a natural for night time displays since people come from all over to see the sun rise and set at the Canyon. With controlled light we could produce a dozen sun rises and sun sets each night. If this proved popular, we could build a dome over the Canyon and light it artificially 24 hours a day.

Maybe we could light the Atlantic Ocean like a large swimming pool, using under water lights in many colors. Imagine the Atlantic any color we wanted.

LOOKING AHEAD

If readers feel that I have proposed future projects too farfetched, or that I have treated serious subjects lightly, I ask that they look back a few years. How many of us would have predicted that the lighting industry would gross billions serving a public eager to jump up and down in dark halls, suffer hearing loss from over amplified sounds, and be surrounded from floor to ceiling with flashing moving lights? Would this prediction have been taken seriously?

In lighting the future, only one thing is certain: we must adapt.

Help Wanted

First published in *Lighting Dimensions*: September/October 1984.

Set/lighting designer/tech director. Ph.D. preferred. Must have extensive professional experience, record of scholarship. To teach technical courses, speech, acting 1, and coach women's hockey team. Design and build 8 shows including 2 operas and 2 musicals. 12-month appointment. Salary $8,102–$9,204.

Mr. Green, it's nice to meet you. I see here that you have just published your second book on design, done four Broadway shows, and been the assistant coach of the women's Olympic hockey team, but you seem very young."

"Well, sir, I got my Ph.D. when I was 19 and that gave me a few extra years to get some experience."

"I must tell you frankly that the committee is impressed with your record, and your portfolio is beautiful, but finds that you have a serious gap in your training. We see no indication of work in speech."

"Sir, I've talked since I was two months old."

Wanted—young stage designer to design theatrical displays for chain of shoe stores. Call 212/154-7652 for appointment.

Before you sit down, I have one question. What newspaper do you read?"

"Um, ahhh, err...the *New York Times*, I guess."

"Sorry. We can't use you. Only someone who reads the *Daily News* will understand our customers. NEXT!"

"I'm sure glad you want to hire me, Mr. Peters. I think I understand how everything will work at the playhouse, what my responsibilities will be, and it looks like it will be hard work but a lot of fun."

"We have a marvelous time here."

"There is one thing, though, that you haven't mentioned."

"What is that, son?"

"You haven't said how much you are going to pay me."

"First of all, son, everyone here is more interested in the work than in pay. I hope that you, too, are as dedicated as the rest of the staff to furthering the art of theatre. We have a mission and a sacred trust to serve the people of this community. But of course, we all have to support ourselves. You will get a bunk in the barn dormitory—FREE. We give our staff 12 meals a week—FREE, and the staff splits the proceeds of the tee-shirt concession. But the most important pay is the chance to show your work to the prominent people who see our shows. We have had designers who were willing to pay us for the opportunity of working here and putting this playhouse on their resumes. We give you a priceless professional advantage."

"Oh. No pay, right?"

"Right."

"It sure is a pleasure meeting you, Mr. Obertine. I gave a paper on your work when I was in school. You have always been my favorite designer. The thought that I may be able to be your assistant is a dream come true. I know I have a lot to learn and working with you would be an invaluable experience. Professor Buttermilk always said that…"

"What have you got to show me?"

"I tried to put a variety of work in my portfolio. If I can only get this zipper open…there. This first one is for a production of *Worker's Blood Oils the Machines* by Fritz. The concept was an evocation of bodily parts filtering down…"

"Where's the drafting?"

"This one is for *Super*, a new comedy by B. Boogie, set in California. We wanted to give the feeling of desert colors washing over tanned torsos submerged in…"

"Where's the drafting?"

101

> Department known for traditional skills
> seeks lighting designer to add computer expe-
> rience to existing programs. Instructor/Asst.
> Prof. Tenure track position.

"We try to be up to date here, young man. I always say that if we don't keep up we're as good as dead. Why, I remember when I was getting my degree at Southern Normal Tech, Edison came to speak. Oh, he was an old man then, of course. He talked about keeping up with the wonders of science and he was an inspiration to us all. I remember that day as if it were yesterday. As I was saying, young man...what is your name again? Oh yes, Bunter. You're not by any chance related to that young fellow Bunter who works with Bobby Jones? I never see Bobby's name in the papers any more, I wonder what he is doing. Oh, you're not. Too bad, that Bunter was a spiffy dresser. Oh well. Now tell me about these computers and what you have been doing."

"Right, sir. You see, I've been a bit bored with the PC and Mac so I'm interfacing hard disk main frames with modifications of HP 150s and TRS 80s by adding modem capabilities that connect power and CAD/CAM possibilities up to 640 RAM. That gives me a 512 x 390 graphics display and plotter capability that with peripherals will knock your socks off. I've set up a data base 16 to increase the megabytes, put in serial ports, and reduced the number of floppies to zip."

"I...see..."

"Hello Myra. I got the show. Yeah, it went great. Well, maybe not perfect, but that's the funny thing. I had my portfolio and he looked at everything, but he didn't seem to know what he was looking at. Yeah. Well wonder if he really cared what I had. I don't know. You see, he looked at everything but didn't ask any questions or anything. He just nodded and said 'Ummm' every now and then. Oh, well, it opens in December. But wait. Let me tell you about what happened. He looked at my stuff in that funny way and said that Jack had spoken highly of my work. I got the idea that it was Jack's recommendation that did the trick. Yeah, you know—it was like I could have shown him manure and it would not have made any difference. No, I don't know. It was like he had made up his mind before I even showed up and the meeting was just a matter of form. Hey, meet me for a drink at Prompter's Box. We'll celebrate."

Dear Tom,

I'll pick you up at the airport. The committee meets at 2:00 so that will give you time to check into the hotel and clean up. The meeting with the Dean and President are the next day and we will have time to show you around in the morning. We will have dinner with a few of the faculty that second night. The third day we can show you around the town and look into some houses that might be suitable for you. There will also be time to talk to the placement people who would know about jobs for Mary. Everyone is anxious to meet you and feels that it would be great if you were on our faculty. I can't make any promises because we have to see at least two other people but it looks good from this end. Best to Mary and the boys.

Sincerely,

Jim

Dear Mr. Tangle:

Thank you for making yourself available to meet our committee. When you arrive at the airport, take the airport bus to the twelfth stop. That will be the Hot Dog Diner. Phone me from there and I will come and pick you up.

We should be here at the campus by 11:15. At 11:20 the committee meets. We will have a quick lunch at the student cafeteria and then at 1:00 you will meet the Dean. At 1:15 you meet the President. You will then have a few minutes to look around the College before the airport bus leaves at 2:20.

We look forward to meeting you.

Sincerely,

James Smith, Head of Search Committee

"Hey, Paul, this is Harry. It's Tuesday at 2:00 p.m. I hope your machine takes long messages. Call when you get back. Hampton has a new show that's going into the Lyric in May. It's an English thing that was done in London three years ago. Maybe you remember it—*Mice*. Not a bad show. About two couples who have a rodent fetish and meet through the personal column in the *Daily Mail*. With a rewrite here and there it could run. Anyway, Donna is doing the rags. Hampton has this kid Michael Bone he wants for sets, and even though I screamed for John, Bone comes with some money from some place. So I insisted on you. Sort of an exchange. I think it is all set. Get your agent to call Hampton. Too bad it won't be all of us together on this one, but the kid is supposed to be good and it is not a tough set. Oh, almost forgot, they signed Rita Toi for the lead. I know she's nuts but if she insists on purple back light to highlight her hair well, you can work that in. And thanks for Thursday night. We had a great time. Call when you get in."

Dear Pierre,

I was overjoyed to learn that you have become Dean of Fine Arts at Small Midwestern College. You deserve the success you have achieved. After the tenure difficulty here at IPU we all felt badly and attempted to institute new guidelines for tenure. Not much has changed, however. And you must be happy to be free of the isolated atmosphere here.

I want to tell you, because I have had no chance before, that I had no idea of your relationship with Inga Schmirborg until after you had left. You may or may not know that she and I were also very friendly at the same time and, in fact, I have since learned that she was also involved with Griggs in Animal Husbandry. She was rather precocious for a sophomore. Did Sally Jane ever know what was going on?

I would not mention this past history were it not that it bears on my present situation. Patty found out about Inga and left me. She was really using this as an excuse because she had been having an affair with Terry McDober in Remedial Philosophy and after his wife joined the Women's Cooperative in Cross Cultural

Physical Examinations, Patty moved in with him. Stella LaRue, our department secretary, married Crackers Finstrum from Australian History after his divorce became final. Cracker's ex-wife, Andrea (a lovely young woman you might remember as a student; her maiden name was Peterbow) and I are planning on getting married in July.

As you can imagine, there are times when working relationships at IPU become strained. It was therefore with great interest that I read the ad for a position at your place for a lighting designer. Many people here would be pleased to write recommendations and...

EDUCATION

[f. L. *educatio*]

The process of
nourishing or
rearing...

●·················

To Whom It...

First published in **Lighting Dimensions**: January 1980.

I am happy to write this recommendation for Mr. C. Clamp because his dedication, sense of purpose, and refusal to give up in the face of great obstacles makes him an outstanding student.

Mr. Clamp has been a student of mine in three lighting courses, and I have seen his work on crews for many college productions. He has shown dependability and even inventiveness, considering that he was often slowed down by braces, casts and bandages. As a measure of his devotion, I have seem him insist on focusing lights with one hand, the other arm and hand strapped in a sling. Others would have been stopped by the various physical difficulties that Clamp has suffered, but even after the first electrical shock experience he vowed to continue.

Mr. Clamp has faithfully attended class when not in the hospital. He has learned a lot about lighting. He no longer tests circuits with his fingers, certainly not the way he did on top of a 20 foot ladder. Nor will he ever again splice live wires while standing barefoot on a metal cat walk. One must classify Mr. Clamp as a person with a charmed life. In fact, some of us on the faculty feel our lives have been made more exciting by this student. A certain expectation, a sense of adventure existed when he was around.

I recommend Mr. Clamp for any lighting job where his single-minded devotion will aid a project, and where hospital benefits are available.

Mr. Bernie Filament has asked me to write a few words about his personal skills and character. One must say that Mr. Filament's most outstanding quality is enterprise. He has shown a remarkable ability to take advantage of every opportunity here at the University. No matter what questions the Dean may have had, Mr. Filament's use of the shop in the early morning hours showed how imagination and ambition can accomplish miracles. Besides producing a complete line of class rings, no one believed he could turn used gel frames into baby potty seats.

There were some questions of propriety when some secondary schools complained that the lamps they were buying from Mr. Filament had already been used at the University. But the Department accepted Bernie's explanation that with his tremendous volume some mistakes were bound to occur.

Mr. Filament has also served the Department in times of emergency. When, just before a major production, all of our ellipsoidal spots were missing, Bernie saved the day by renting us the very instruments we needed and at a very reasonable price.

Since the University cannot pay students for crew work, we have at times had difficulties competing with the wages Bernie offers students for working on his local jobs. Hardly any occasion in this area is without Filament's lighting. We have all wondered how he convinced families to use theatrical lighting for a circumcision, a Thanksgiving dinner, and a tax audit.

Bernie Filament is the kind of student for whom the faculty can predict success without the slightest doubt. I offer this recommendation freely. That Bernie somehow now holds the mortgage on my house has nothing to do with the praise I write.

Ms. Betty Panel has my highest recommendation. She is one of our most attractive students and has outstanding qualities that are evident to all. It is difficult to speak too highly of her warm personality, her willingness to give of her self, and her concern for the needs of others.

Ms. Panel's caring nature is matched by her sense of discretion. Furthermore, she takes precautions to see that no unpleasant surprises complicate her personal relationships.

She is remarkably open and willing to experiment. She is an eager student, always ready to learn. Because she is a good student, she also has some things to teach the faculty. Ms. Panel's presence was felt through out the Department and she was affectionately known as "Patch."

I can warmly recommend her for whatever position you want to put her in.

Big Ole Stepper Lens has laid it on me to script a recom bit for the man himself. No sweat. He is an outrageous, laid-back dude. The whole bad bunch here at Southern Surfer U. dig his funky cool.

What Ole Stepper does with those strobes and flashers would blow your mind. We don't get our s— together very often to do real plays, but Big Step has put down the sparkles on messes of our nightly jams. And the foxes hang on this stud when he hits the buttons. But Ole S. L. is cool and ignores the broads.

Sometimes he really flips and is out of it. The doc jaws about seizures from strobe pulse. We all think he is just blown; you know what I mean.

Ole Step hangs in there about as nasty as anyone I know. He is truly a bad dude, top of the class.

Phillip Fader has been a student of mine for four terms. The classes he has attended were designed by me to give a basic education in lighting design by providing both excellent theoretical work and practical experience. The students who have graduated from my courses have an extremely high degree of success in the professional and educational worlds.

I start the students out with terminology and aesthetics. They use the text, *The Wonder and Beauty of Light* which I wrote in 1961 and which has gone through four printings. This is a particularly useful book for students (and has been praised in all the best journals).

My course in lighting theory uses my spiral bound monograph, *Forty-Seven Ways to Better Lighting*, which I publish myself. It is available by mail order from the above address. Most students benefit greatly from the wisdom of this short volume.

In practical courses I insist that students work crews on productions for which I do the lighting to observe the best possible techniques.

At the end of their training, some students are allowed to work as my assistants. I have to select only the best to help. The experience of working with me has been of enormous help to those lucky enough to have been chosen.

As you can see, Phillip Fader, having been my student and assistant, is highly qualified and recommended.

OXYMORON, R. I. P.

First published in *Lighting Dimensions*: September/October 1985.

Professor of English S. Whitherford Jones is fond of dramatizing his philosophical concerns. I ran into Jones at the book store where he was flipping through a copy of *How I Made a Million Dollars by Writing Better Paragraphs*. He was wearing a black arm band.

"My condolences, Jones. Who died?"

"Oh, you noticed my arm band? Most people pretend they don't notice because they don't want to hear a tale of suffering and loss."

"To tell the truth, I don't want to either. I thought it the decent thing to inquire."

"Not to worry. It's nothing like that. I'm wearing the arm band to make a point. Let's have a cup of coffee and I'll tell you about it."

At the Commons cafeteria, he continued. "Oxymora are dead. They are the last to go. Nothing is left. First metaphors, then ambivalence, then ambiguity. The language is a wasteland. There is no playfulness, no fun, no truth. I'm in mourning for all the figures of speech. Have you tried a metaphor on your students lately? They screw up their faces and snarl, 'What's that supposed to mean? Tell us in straight language what you are trying to say.' Heavens, man, I am destroyed. I am in mourning for an age when important things could be communicated through language, and I don't mean body language. But maybe your students are different.

They're artists. They must have some poetry in their souls."

"I'm afraid I see the same problem in my students. They don't seem to understand scripts. Give them a play with poetry or characters who are contradictory, and therefore human, and they fall apart."

"My condolences to you. We're in this together. Do you want to wear my arm band for a while?"

"No thanks, I'll just rend my garments and pour ashes on my head."

112

"You theatre people are more dramatic. I wish I could afford the mending and cleaning bills. But you make a point. The torn clothes and ashes thing is a metaphor that expresses a depth of feeling that saying 'I'm sad' can never do. The only trouble is that my students are afraid of strong images and when they write about what they have read or even about what has happened to them, they write a string of facts and use the most common clichés."

"Design students do the same thing in a different way. They never get past the basic action of a play. They see the whole thing in terms of who moves where. It is like a baseball game without the strategy. Every play becomes a TV sitcom; it has action, one-dimensional characters, and an easy solution at the end. That way of thinking makes *Hamlet* a melodrama that any four-year-old can understand. Students are afraid of mystery. They want answers but only on the simplest of terms. They can't understand that questions can be answered by other questions. They are frightened by ambiguity."

I was getting excited as I remembered past frustrations, times when my ability to explain a play fell short.

Jones calmly nodded at what I was saying. "Tropophobia," he said. "The poet, Donald Hall, calls this phenomenon by a neologism, tropophobia, 'the fear and loathing of metaphor.' He claims that it stems from the fear of being taken in by language. You know, the president calls a missile the 'Peacekeeper.' Coke is 'The Real Thing,' and then changes its formula. Everyone is wary of loaded language. I guess if one thinks that language is suspect, the only way out is to opt for action. But the funny thing is that when these kids communicate they talk to each other in language that is entirely ambiguous. I don't know what they mean." Jones did an imitation of a couple of students:

"I just got a B in psych."

"That's cool."

"I just won 10 million in the lottery."

"That's cool."

"And then try to get anyone to define 'cool.'"

"Maybe you have the answer. I'll just tell my class that *Hamlet is* cool and they'll know what I'm talking about."

"Sure, and *Macbeth is* gnarly."

"That would make things easier, but it doesn't solve the problem of tropophobia. What do we do?"

"I keep plugging away and giving examples they may remember years from now. In any case, I love the examples and use them because they're fun for me." I didn't have to ask him to explain.

Show me a professor who is not primed to repeat what he tells a class and I'll show you a dead professor.

"I take a description of an event and show how a poet uses language to make it touch our souls. Let me try to remember Plutarch's words."

(I knew damn well he had them memorized.)

...she came sailing up the river Cydnus, in a barge with gilded stern and outspread sails of purple, while oars of sliver beat time to the music of flutes and fifes and harps...The perfumes diffused themselves from the vessel to the shore, which was covered with multitudes, part following the galley up the river on either bank, part running out of the city to see the sight. The marketplace was quite emptied, and Anthony at last was left alone sitting upon the tribunal.

"Here is what that glorious thief (pardon my oxymoron) Shakespeare does with the same thing:"

The barge she sat in, like a burnished throne,
Burned on the water. The poop was beaten gold,
Purple the sails, and so perfumed that
The winds were lovesick with them. The oars were silver,
Which to the tune of flutes kept stroke and made
The water which they beat to follow faster,
As amorous of their strokes...
...The city cast
Her people out upon her. And Anthony,
Enthroned i' the market place, did sit alone,
Whistling to the air, which, but for vacancy,
Had gone to gaze on Cleopatra too,
And made a gap in nature...

Jones recited the poetry with passion and emphasized the figures of speech. Like many English teachers, he had a bit of ham in him. And he was right, without the metaphors and the poetry, the story becomes Dallas in togas.

I also recognized the glazed look he had developed and knew that I was in for the rest of whatever lecture he was recalling. I can't remember it all, but a few disconnected quotes stayed with me.

"...our most serious talk about the nature of reality is always and inevitably metaphorical..."

"...a poetic mind is purely and simply a syntax of metaphors."

"Everything is only a metaphor; there is only poetry."

"Every original metaphor contains a submerged riddle."

"Metaphor focuses both thought and emotion."

114

Walking back to the Drama Building, that large metaphor for education in theatre, a striking coed passed by "...so perfumed that the winds were lovesick..." I tried to keep my mind on figures of speech.

Teaching Design In A World Without Design

First published in *Theatre Design & Technology*: Winter 1989.

Those of us who attempt to teach theatrical design sometimes mimic our own teachers and sometimes devise new methods to fill the fissures opened up by changing circumstances. Aesthetic, political, moral, economic, philosophical, technological, and social changes should indeed alter what we teach, the way we teach, and maybe even why we teach. It is a crazy world out there and we are obligated to make sense of it for our students, neither simplifying it to the point where we lie, nor mystifying it to the point where we are unhelpful.

Teaching design is, as Garrison Keillor says of growing old, not for the faint of heart. We are, however, timid of mind and lazy of soul, and have retreated to the comfort of teaching technique rather than ideas. Having separated ourselves and our students from the sources of theatrical impulse, we grumble about the educational system.

Professors like to complain about how ill-prepared their students are. Kids today don't know history, literature, or art. They can't analyze a play or write a clear sentence. We blame any number of recent social changes—and yet, I wonder if what we say was not said about us? Has any generation of students satisfied its professors? For my own part, I blush in memory of what I didn't know, and grieve for the vastness of my present ignorance.

But ignorance itself is not a sin. I worry about our student's attitudes toward their ignorance: their lack of curiosity, their artistic conservatism, their political and social neutrality, their inability to take daring leaps of imagination. And I am thinking not only of the students at my own university. For several years, at portfolio reviews and at USITT student exhibitions, I've noted that almost everything shown could have been designed thirty years ago.

How do you explain the paradox of students raised in an age of music videos, concert tours, and hyperactive commercials creating designs that are tepid and dull, with none of the verve of their own environment? Is it possible that our students lack the analytical skills to isolate the components of the stimuli that engulf them? If so, they are powerless to employ the strongest tools of their times.

Perhaps their timidity is a response to a world gone mad with images and irrational stimulation. Surrounded by this confusion, they retreat to an art that has form, rules, and reality. Although one role of art is to create order out of chaos, it is not useful to create an order that was only valid thirty years ago.

If, in fact, our students are retreating to safe ground, their conservatism may well spring from a profound fear of the future. That we have so few risk-takers and so few idealists among the young may be a reflection of a fearful society that is itself looking for answers in the past, a time that was more comprehensible and comfortable. The university articulates those concerns that are operative but incoherent in the larger world. It has reacted with a debate. The deceptively tranquil academic sea has been roiled in the wake of Allan Bloom's book, *The Closing of the American Mind.*[1] Smaller waves were made by E. D. Hirsch, Jr. and his *Cultural Literacy,*[2] and by Russell Jacoby's *The Last Intellectuals.*[3] Bobbing to the surface are articles and critiques of these books and what they say about the health of the American university.

"It is our students' lack of a liberal arts background and the tools it provides which drives teachers of theatre to despair."

It is therefore no surprise to see articles which discuss the state of theatre design education as it relates to the more general controversies in academe. Look at "What Went Wrong in Stage Design Training" by John Ezell and Felicia Londre[4] along with a response from John Gleason, "Another Voice,"[5] and several articles in Volume 31 of the *Performance Arts Journal*, especially philosopher Bruce Wilshire's piece on the place of drama education in the university.

Each of the critical works, from Bloom's absolutism to Gleason's laissez faire attitude, points to a paradox. While professional training depends on critical and logical thinking as well as the means to express and communicate ideas, those very abilities are best learned and honed in liberal arts studies. It is our students' lack of a liberal arts background and the tools it provides which drives teachers of theatre to despair.

Theoretically, remedial work in the liberal arts can correct the deficiencies we find in our students. This is, in fact, the remedy suggested in many articles, including the one by Ezell and Londre.

117

We find, however, that students resist liberal arts studies either from an ignorance of their value or, more likely, from an antipathy fostered in them by our society, which is rejecting the ethos of the traditional university.

Ideally, the university is a bastion of reason where truth is pursued through intellectual exploration, the contest of ideas against ideas, and a value system which rationally judges one idea superior to another. Even in the arts, accepting the importance of emotion and inspiration, values are arrived at through rational standards. This difficult task of intellectually explaining our emotional responses is what makes teaching in the arts the challenge that it is.

Many in our world today mistakenly see only two possible ways of knowing: objective truth, which is scientifically provable; and subjective truth, which is emotional, poetic, and completely relative. I share Bloom's criticism of an era in which students find one idea as good as another and defend each emotional response as a private matter immune from any value judgment.

This relativist approach leaves me nothing to teach but technique. I see in it no hope for beneficial change. It closes avenues of communication because we cannot compare responses. In short, it is antithetical to the concept of a university. A tension is created between the act of teaching and the act of understanding when so many of our students trust only their own emotional reactions as indicators of truth.

Students must find it difficult not be relativists. When our former president's schedule is determined by an astrologer, it is a frightening sign that rationality is losing ground. When religious fundamentalists sue to have creationism taught as a scientific theory along with evolution, it is evidence that the difference between theology and science is not understood. And when humanism, a major part of a university education, is made a term of contempt, it is clear that anti-rational forces have gained power in our society.

The complex reasons behind this tilt toward the irrational include the seductive power of television images, a growing distrust of the meaning of words, and the frightening effects of new technological and scientific discoveries.

It may be that our students' fear of the future and our own academic malaise are reflections of a world that is still digesting an overdose of scientific discoveries. In this period, it is difficult to remain sanguine when we are faced with the possibility of total nuclear annihilation, death from pollution, an advanced communication system which says less and less, computers which seem about to replace the human brain, and biological science which is cloning cells and producing new strains of potentially dangerous viruses. Reactions to such developments have resulted in a society that does not trust man's ability to make a better world through reason.

Henry Fairlie says that our reaction against technology began with the Vietnam War:

> ...technology, it was observed, not only fouled the environment, but had proved incapable of winning a war against guerrillas in the jungle. And beyond this, of course, has been the belief that science itself has somehow betrayed us, that it promises evil and not beneficence.[6]

When science, the rational objective way of knowing, betrays us, it is tempting to turn to the emotional and subjective, as many students have done. As teachers, however, we cannot join them. Moreover, anti-rationalism is coupled to cynicism: there are no heroes, no values, no hope for a better future. And as teachers, we cannot share this view. Risking a broken heart, we stand for optimism against despair, and, against heavy odds, for the perfectibility of mankind.

Taking the long view, we may find comfort in the cyclical nature of civilization. We must remember that each great time of scientific discovery has appeared to move man one rung lower from the gods: after Galileo, the universe no longer revolved around the earth; after Darwin, the apes were our cousins; after Freud, the conscious mind was not the sole master of our actions or thoughts. There is an understandable resistance to each seeming diminution of man's stature, but eventually, there has been a renewed sense of wonder, curiosity, and rationality which is able to encompass the added complexities of existence.

The present situation has a special importance for those of us teaching in the arts. Looking at the great periods of Western artistic accomplishment—the Greeks, the Renaissance, the Modern Age—the arts appear to flower in societies which have confidence that human beings can use logic and reason to solve problems. It is in those high points where reason and humanism hold sway that we have high points of culture. This is not to say that great art is limited to these times; the need to create is so powerful that it operates in any atmosphere. But in our Western culture, art is the natural partner of humanism and an accepted balance to rationality.

The arts play a special role in a society such as ours, which is founded on a linear rationality and humanism. They offer a permissible contact with the irrational, the emotional, and the myste-

119

rious forces that logic cannot explain. The arts are permissible because they are created by humans and therefore add to the glorification of mankind. In other words, the arts become, oxymoronically, a secular religion in a humanistic society.

Obviously, the arts have a problem in the present anti-rational, anti-humanistic climate. This problem may be temporary. When we recover our faith in reason, all will be well in the arts. Our present dysfunction, however, gives us a chance to look at why we go through periodic ups and downs.

In Myth, Literature and the African World, Wole Soyinka says that European culture, divorced from a cosmic connection, has operated in a series of "intellectual spasms," moving from one fashionable style to another, while African theatre has been a "communal evolution of the dramatic mode of expression."[7]

All art, all theatre started with a cosmic connection. Humans searched for their place in a natural world that was powerful beyond knowing. Soyinka says:

> In Asian and European antiquity, therefore, man did, like the African, exist within a cosmic totality, did possess a consciousness in which his own earth being, his gravity-bound apprehension of self, was inseparable from the entire cosmic phenomenon.[8]

"I believe that art which fulfills the mythic role, the ancient mission of art, could be the antidote for the fear and disjunction of our students."

We can make a distinction between those works which retain a link to art's original intent and those which do not. Art that tries to explain and control ineffable forces; that joins individuals to the group and the group to ineffable mysteries; that forces us to consider our mortality, our ideals, our morality in terms of the eternal: this is art that connects us to the past, present, and future, to the natural world, and to ourselves.

In contrast, we have works designed to make us forget our problems, weakness, and smallness, works that lull us, that ignore our mortality and the universe. I am not dividing art into tragedy and comedy, for the best comedy rests in the first category while much of today's "serious" art focuses on the disconnected exploration of an individual's psyche or petty interpersonal relationships.

In The Hero with a Thousand Faces, Joseph Campbell gives us another way of looking at our art by contrasting fairy tales and myths. In a fairy tale, the hero goes through the same trials in his quest as does the mythic hero, but in the end, the benefits gained redound to the individual hero. In a myth, the hero's victory nourishes the whole community.[9] Likewise in theatre, we can separate those works which speak to a universal need and those which position us as voyeurs to a particular and mostly diseased situation.

I believe that art which fulfills the mythic role, the ancient mission of art, could be the antidote for the fear and disjunction of our students. Such art still exists outside our own Western, European, white culture and it is both foolish and dangerous to ignore it. Director Peter Sellars, writing in the *Los Angeles Times*, says,

> ...this may be the time to notice that we are surrounded by a range of cultures far more ancient, distinguished, and profound than our own. My generation may be the first American generation that has had to notice that it's not a white world.[10]

Sellars characterizes non-Western cultures as:

> ...centered in participation, not spectating...they are not susceptible to commercial exploitation. They are about sharing. They are about a graceful attempt to acknowledge and respect the unknown. They are about boundaries and mutual obligations. They are about friendship. They are about value before profit.[11]

Many academics share Sellars's views and have attacked Bloom for excluding women and non-white cultures from the basic education he proposes. Supporters of Bloom say that, if there were works by women or non-Western authors equal in stature to the great Western classics, those should be taught, but such works don't exist.

I'd like to straddle the fence and suggest that while there are not the same kind of classic works in non-Western cultures (because those cultures were, for one thing, not as egocentric) there are, however, attitudes, thought processes, and cosmic interpretations that we must study and, ultimately, incorporate into our way of looking at the universe. In other words, our study of other cultures will be done, not from single great works, even though there are some, but mainly from sympathetic anthropology, history, ethnology, criticism, and, as much as possible, from visiting artists and foreign travel.

The end result will not be that we can reproduce an African or Asian ritual and fully appreciate its nuances and flavor, but that we can use our study to enrich and inform interpretations of our own classics and new works which speak to a new world culture.

This is hardly an original idea. Peter Brook and Ariane Mnouchkine are Western theatre artists drawing on other cultures. Going in the other direction, Wole Soyinka and Tadashi Suzuki use Western ideas. In the past, Martha Graham, Noguchi, Picasso, Gauguin, and countless others have looked outside of their own cultures for inspiration (although they occasionally have missed the point, according to Eileen Blumenthal).[12]

Our design training has mostly ignored this vital rich vein to be mined, but we are obligated to free our students from the cultural chauvinism that binds them to a kitchen-sink realism. They need to see the power and beauty in ritual. Because Western theatre has become disconnected from religious ritual, designers, of all people, should master the use of ceremony.

Design students, concerned with movement and space, should understand that Western concepts of time are not universal; indeed, they can be a stumbling block to comprehending foreign works which are based on other temporal metaphors. (For example, George Lakoff and Mark Johnson in *Metaphors We Live By* argue that we think of time not only as space but as a valuable object to be saved or spent. They say that the Westernization of other cultures means the imposition of the metaphor "time is money."[13])

Our basic ideas of time are lineal. The future is ahead, the past is behind, and as the beer commercial says, we only pass this way once. In general, time in Asia is circular; we live, we die, we live again, we die again. In Africa, time is simultaneous: the unborn, the living, and the dead all exist at the same time, which is one up on Robert Edmond Jones's vision of representing conscious and unconscious worlds at the same time.

There are, I think, two keys to unlocking the mysteries of other cultures for our students. One is the understanding that Campbell gives us that myths particular to one people are, in fact, universal. Stories of the creation of the world, floods that wipe out evil, virgin births, the trials of heroes—all of these appear in almost all parts of the world in strikingly similar forms. If our students can see that we are related to other cultures by our mutual questions, longings, and fears, they have a better chance of stretching their imaginations by relating to what now looks exotic and irrelevant.

The other key is an understanding of metaphor as an expression of truth distinct from linear scientific logic and emotional relativism. Our students need to see that metaphors express a reality that is both more accurate and more mysterious than either objective or subjective knowledge. Poetic metaphors link us to other cultures through a common language of the soul, a language that is psychologically and spiritually more real than reason.

Theatre itself is a metaphor. Sets, costumes, and lights are metaphors. Our art lies in our ability to hone and polish metaphors that communicate to an audience. Our metaphoric messages must be precise and poetically mysterious.

Theatre was born because the world often seemed without design. Our ancestors gathered together to hear tales of gods and heroes. Through ritual and ceremony, they sought to make sense of chaos to control and appease forces too grand to be imagined, to chart their relationships with nature and each other, and to find comfort as members of a group.

Not much has changed. We and our students need theatre for the same reasons today. More than just settling for studies of an ephemeral technology, it is time we teachers re-kindled the fire that theatre stole from the gods so long ago.

ENDNOTES

1   Allan Bloom, *The Closing of the American Mind* (New York: Simon and Schuster, 1987).

2   E. D. Hirsch, Jr., *Cultural Literacy: What Every American Needs to Know* (Boston: Houghton Mifflin, 1987).

3   Russell Jacoby, *The Last Intellectuals: American Culture in the Age of Academe* (New York: Basic Books, 1987).

4   John Ezell and Felicia Londre, "What Went Wrong in Stage Design Training," *Theatre Crafts,* October 1988, p. 42.

5   John Gleason, "Another Voice," *Theatre Crafts,* October 1988, p. 42.

6   Henry Fairlie, "Fear of Living," *New Republic,* 23 January 1989, p. 16.

7   Wole Soyinka, *Myth, Literature and the African World* (Cambridge: Cambridge University Press, 1976), 38.

8   Soyinka, 3.

9   Joseph Campbell, *The Hero with a Thousand Faces* (Princeton, NJ: Princeton University Press, 1968), 187.

10 Peter Sellars, *Los Angeles Times,* 5 February 1989, calendar section, p. 3.

11 Sellars 3.

12 Eileen Blumenthal, "West Meets East Meets West," *American Theatre,* January 1987, p. 11.

13 George Lakoff and Mark Johnson, *Metaphors We Live By* (Chicago: University of Chicago Press, 1980), 7.

DISSERTATION

First published in **Lighting Dimensions**: January/February 1985.

"Let me get this straight. You want to write a paper or something about me and my lighting design? You look pretty old to be writing school papers. How old are you?"

"Well, I'm 36. But you see it's not exactly a paper. I'm writing a dissertation. which is necessary to get my Ph.D."

"Hey, I'm only 33 and you're writing a paper about me? That's intense."

"It's not just a paper…"

"Whatever. I'm freaking. Hey, can I have some copies when you're done? Maybe two, three hundred to pass out to friends? They would be blown away."

"It's not really a paper. It's a dissertation, and it's a book-length work. It would be kind of hard to give you so many copies. Usually there are just a few, one for the library, one for me, and one for the department."

"No comprendo. You write a whole book and then nobody reads it? Man, that's weird "

"It's sort of like a practice. Yes, that's it. If I were a musician, I would do scales to warm up and it's not important if anybody hears them. The dissertation is to warm up for writing other things later on."

"What are you warming up to write?"

"Oh, most of us never write anything again."

"Wow. You're on some bad stuff."

"Let me just say that I'll lose my job next year if I don't finish this thing. I'm an Assistant Professor at Prestige University and they hired me with the understanding that I would finish my Ph.D. by next year or be out on my tail."

"It's like an initiation? Or something like a union card?"

"You might look at it that way."

"Okay—but why are you doing this thing on me? I've only been doing lights for four years. There must be guys around who've done more stuff to write about. I'm good. Lordy, I'm a special Joe. But can you fill up a book about me? I always thought that you guys wrote about dead people. You got a bunch of letters and programs and old stuff to write about. That's why I never read those things."

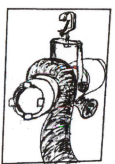

"That is the way it is often done, or was done in the past. But according to Professor Katzenmeyer—he's my professor for the thesis—the good subjects have all been done. Oh, I don't mean it like that. I mean that the old guys, the ones from history, have all been done. There is nothing more to say about historical figures so we have to do our work on living artists."

"Now wait a minute. I didn't just fall off the watermelon truck. Are you telling me that there are more of you Ph.D. fellows around than there are artists from the past to write about?"

"I know it must look funny from the outside, but there was not much written about theatre artists—that is, designers—in the past. That makes it tough for scholars to dig up people and stuff to write about. There is definitely a shortage of material, especially with so many people doing graduate work."

"So you came up with little ol' me to write about."

"Well, Scnubler from Mex Tech is doing Jim Moody and Betty Noo from Florida Water and Art is doing Bob Brand…"

"I get it. I'm third choice."

"To be honest, about seventh."

"I shouldn't have asked."

"It really sounds less flattering than it is. Professor Katzenmeyer contends that the younger designers, while having less work to talk about, have a clearer memory of why they did certain things. You were my number one first choice because you have done so little."

"Oh."

"I've put my foot in it again. I didn't mean that the way it sounded. Your work is terribly important. Even though you haven't been working a long time, those things you have done have caused quite a stir and have been terribly influential. Who can forget those twelve girls in *Lets Get Lit*, laid over lighted fluorescent tubes…I mean, lying across fluorescent tubes? Or the dachshund wrapped in chase lights being led through fog in *Long Dog's Journey?* As long as I live, I'll never forget the color changes you did for the Emetics Victory Tour at the moment they dismembered those three babies."

"Man. you must have seen everything I did."

"I try to keep up. But I did miss that show in Detroit, the one where you did the whole thing in Christmas tree lights and X rays."

"Too bad, that was my best. But it wasn't X rays, really. We had portable fluoroscopes on all the actors so that the audience could see their real visceral reactions. Talk about revealing yourself! Not only was it a treat for the audience; we discovered one polyp and a garnet ring missing for twelve years."

"But wasn't all that radiation harmful?"

"Yeah. But actors will do anything to get a part. Tell 'em to cut off their hair, take off their clothes, lose twenty pounds…they'll do anything. Funny thing was the EPA finally came in and made us stop because the audience was getting too much. The meters made Three Mile Island look like jelly beans in comparison."

"It sounds as if you don't have much respect for actors."

"I never thought much about that, but maybe you're right. They always get in the way somehow. And when you want them to be in a certain place, to get in a light, they always mess up. Come to think of it, they are always complaining, too. The color isn't flattering, the stage is too dark, there's too much light on the projections and not enough on them. They are a pain."

"Do you prefer musicians?"

"Sure, they're a sweet group. They like it when lots of things are going on all around them. The more fancy stuff, the better. Smoke and fire, mirrors. flashing things…those cats can appreciate lighting."

"But doesn't all of that distract from what they are trying to say?"

"Trying to say? They're not trying to say anything. Hey man, it's not saying, it's feeling. Feeling is what it's all about."

"And you're not trying to say anything with your lighting?"

"Listen up. It's feeling. What should I be saying? I don't have any message. But I can get a lot of people to feel things."

"What do they feel?"

"How should I know? Feeling is feeling. Everybody feels things in a different way. Okay, look. I know what you're getting at. I'm not a complete bozo. You guys in the school talk a lot about communication, collaboration, messages—all very intellectual. That stuff is four-day-old baked Alaska. The real world doesn't care a used Kleenex about that."

"I'm confused. If you don't have a message, how do you know what people feel about what you do? How do you know if you are doing a good job?"

"How do I know how well I'm doing? Man, if those kids who paid forty bucks are screaming and jumping around I know I've done it. And if the boss hires me again, I know. You don't seem to get the picture. Nobody today has any way of knowing if he does a good job except by the emotional reaction of other people. Like, nobody comes up and says, 'That was a terrific job you did.' No. They come up and say, 'I love it.' You see? It's not what I do that's important to them, it's their own reaction to what I do."

"But I read that article about you in the *New York Times* Sunday section. It said that you were trying to demonstrate…wait a minute, I have the clipping right here. It says, 'He represents a new generation of artists whose prime concerns are explorations of a technological world gone berserk and understandable only through a teleological feat of cosmic understanding.' There. I must say, Professor Katzenmeyer was very impressed with that article."

"Maybe Professor Katzenmeyer will explain to me what it means."

"You know, I think I should have another talk with the professor and see where this thing is going. I'm not at all sure that I'm on course. You have been very helpful. I can't thank you enough. I hope I can talk to you again after I work out a direction for my thesis. Thank you. Thank you."

"Sure. It's the door over there, where the laser is writing, 'Out.'"

Ben And The Art Of Teaching Deisgn

First published in **Lighting Dimensions**: July/August 1985.

Two recent phone conversations appear to be connected in a way that may be of interest to readers of this column. The first went something like this:

"Is this the person who writes stuff for *Lighting Dimensions?*"

"Yes. Who is calling?"

"Just call me Ben. I don't want my whole name turning up in print, so as some nitwit somewhere can get back at me."

"Fine with me. What can I do for you?"

"I want you to write about how shitty the teaching is in theatre design courses and how it messes up the kids."

"It sounds as if you have some personal gripe."

"You're damned right I do. I got messed up good. Here I am, after seven years of college and graduate school, teaching at a backwater school and doing the same lousy things to kids that were done to me. I tried for almost a year to get some professional work and after nearly starving I took what I could get. My teachers lied to me. Told me I would be prepared to design any place. You know what? I'm doing the same thing to students. And why? Because I'm selling a service and want a lot of people to buy it, want as many people to enroll in classes as I can get. If I don't, I won't even have this job."

"Wait a minute. You may be a special case. There must be some good training programs around. Hell, I teach too, and I don't think I could look in the mirror if I were lying to all those students."

"I don't want to be rude, but you're a flaming idiot. I got proof that none of these things are worth a damn. Have you seen the new book on scene design?"

"Do you mean Arnold Aronson's book, *American Set Design?* Yes, I've read it. It's a great book. It gives a lot of information and it's fun to read. It's one of the few new books that really covers things that no one else has written about."

"Yeah, that's the one. That's what got me thinking. As I read it, I got angrier and angrier. I was furious when I finished. No, I wasn't jealous of all those successful designers…well, maybe a little. But what got me hopping mad was to see what kind of training the most successful designers had. I made a list and counted up where they had gotten their educations and what they had studied."

"That sounds like an interesting survey. What did you find?"

"Okay. There are eleven designers covered. How many of those do you think had undergraduate theatre training?"

"I don't know. How many?"

"Get this. Out of eleven, two had full undergraduate training, three had a little, and six had none. Interesting? More than half of the most successful designers did not take undergraduate courses in theatre."

"Really? What did they take?"

"Well, there were a couple of English majors, some art history, a couple of art students. And listen to this: the one who did the whole theatre education bit, undergrad plus grad schools, is quoted as saying that traveling around the world for five years would have been as helpful."

"Would you mind if I wrote about this?"

"Hell, why do you think I'm calling you? I told you I want it all written up. But there's more. Four of the eleven dropped out of schools. And those were big-name schools. Just up and quit because they weren't getting what they wanted. And they did it after a short time. They must have seen through the bull real fast."

"What do you make of that?"

"Seems like those people were smarter than I was. Maybe that's why they're successful. If they knew a school was bad for them, or maybe just bad, they also probably know more about lots of other things. Maybe they know more about what plays mean or what art means. No one taught me things like that and I'm not able to teach my students. Maybe no one can teach students that. Maybe that's what talent means."

"So what's the answer?"

"That's why I'm so damned mad. I'm frustrated. If I tell my students to leave school and travel the world for five years, I'm out of a job. But that's just what I'd like to tell them. Or maybe I'd tell them to go and work with Ming Cho Lee."

"Why Ming?"

"Because if you read the book, six of the other ten designers were either his students or his assistants. That's some record. He turns out to be more important than all the schools in the country put together. I'm envious of all those designers who worked with Ming Cho Lee."

"Do you want to know something? I've talked to Ming about training designers and he's as frustrated as you are. There is something going on today that has produced a sort of crisis in teaching design."

"Oh my God. If Ming is frustrated, what can I do? I don't know if that information makes me feel better or worse."

I thanked Ben for his thoughts and suggested he call again with more ideas. It was not more than two days later that I received another call, this time from an old friend. We talk now and again and see each other at conventions. Sometimes we blow off steam by using each other as a sympathetic ear.

"How you doin', old fart? Did you see Doonesbury today? It's one of his college strips. He shows a college president addressing the graduating class. Wait a minute...I'll read part of it. The president says, 'Members of the Class of '85, you have spent the last four years in a trance. Taking your cue from the highest office in the land, you have somnambulated unfeelingly through a wounded world, pausing only to debate the central issue of your era—student parking. Paradoxes abound. By accepting unquestioningly all you have been taught, you have learned nothing of value. As you have enriched your prospects, you have impoverished your souls.' It goes on a little more, but that's the great part. I want to make Garry Trudeau president of the world."

"You called me to read a comic strip?"

"I knew you'd be appreciative, you old bastard. It's only that Trudeau touches a nerve more often than not. I've been stewing about my students for so long that any indication that someone else sees the problem makes me feel better."

"Come on, your colleagues must talk about what has happened to students today."

"Sure, they talk about the problem, but do you know what they do about it? Here we have a student population that accepts everything they're told without question; they only want to be told how to do something. They're not interested in ideas—only practical techniques. They think we can tell them how to create a work of art and then they can go and do it. So what does our faculty do to improve the situation? They take polls of what the students want. Here we have a faculty tearing its collective hair, if that's grammatically possible, saying, 'Where have we gone wrong? Oh please, students, tell us how we can give you what you need to be rich and famous.' The students, who have never wanted to be burdened with the pain of having ideas, answer back, 'Give us more courses that teach us how to do things. We want six courses in makeup. We want hundreds of credits for putting plugs on cables in labs. We want courses in preparing our portfolios."

"Sounds as if you had a bad week."

"No, it wasn't too bad. Found out that the kids don't need orthodontia and got a nice royalty check from the book. But it drives me batty when I see professors being as foolish as they are. Take a poll, take a poll, as if ideas were the product of a democratic process. Heaven save us from polls! Now that I think of it, hell is probably filling out questionnaires. But enough of that. How are you?"

"I'm glad you called. I've also been disturbed about the problems of teaching theatrical design and I've been taking a poll of..."

EDUCATION AT LAST

· ·●

First published in *Lighting Dimensions*: September 1978.

POST GRADUATE EDUCATION IN THEATRE TECHNOLOGY THE UNIVERSITY OF SOUTHERN FORTY-SECOND STREET

A proliferation of theatre schools and drama departments since the Second World War has dumped into the marketplace masses of graduates trained in all areas of theatre technology. None of these graduates, however, has received formal training in those subjects which truly prepare them to function in the professional world. The University of Southern Forty-Second Street is now offering post graduate courses designed to supplement a diet of theatre courses deficient in practical vitamins and minerals.

Coping 104
5 hours 3 credits

The basic course is designed to help the student get his own way in dealing with colleagues. Starting with positive techniques such as flattery (sometimes known as ass kissing), it moves into giving and getting sexual favors, voice training for screaming, when and how to cry, and finally to negative techniques of law suits, both real and threatened.

Guest lectures are given by Supreme Court justices, several residents of the Forty-Second Street area, and a four-year-old tantrum expert.

Coping 206
3 hours 3 credits

This course, titled "Looking Good," covers the important area of outward appearances. The student is encouraged to design a persona for himself, taking various forms of either "sloppy-genuine" or "elegant-a-la-mode". Some time is spent discussing the proper instruments to carry to look professional and efficient. Training is given in the ostentatious use of light meters, slide rules, calculators, scale rules, and view lenses.

Coping 207
3 hours 3 credits

The second term of "Looking Good" is devoted solely to the use of an entourage. Concrete examples are presented by various experts and their assistants (often known as foot pads). This course will prepare the student to recognize when to appear himself, when to send an assistant, or when to appear with assistants and how many.

Coping 307
3 hours 3 credits

The advanced course in this series deals with the extra-physical process of "Looking Good". Starting with techniques of getting publicity without advertising, it leads the student to discover ways of starting rumors which aid his career. The course ends with subtle ways of bad-mouthing the competition. As was wisely said, "It is not enough to succeed yourself; your best friend must fail."

Coping 308
2 hours 2 credits

The most advanced course in Coping involves techniques of getting others to do your work. The expert in this area will know ways of taking credit for the work and ideas of others.

Early Years 116
6 hours 3 credits

For the student just starting his career, practical techniques of solving real problems are explored. Going further than just an attractive resume and portfolio, this course aids the young professional in beefing up his resume with non-checkable items and shows how to write letters of recommendation from deceased celebrities.

Early Years 117
2 hours 1 credit

After the portfolios and resumes are ready, weekly field trips give the student experience in how to carry a portfolio in a high wind, how to get on and off a crowded bus or subway with a portfolio. Advanced work at the end of the semester covers transporting models in taxis and through rain storms.

Support Systems 223
6 hours 6 credits

The basic course in support systems is a key to a successful career. It explores ways of providing income while looking for work. (If a student is fortunate enough to have wealthy parents and has convinced them to be generous, this requirement can be waived on presentation of an up to date bank book.) The major emphasis of the term is on finding a rich mate. The procedures learned include: frequenting the watering holes of the rich; how to tell a wealthy man/woman from a door to door salesman/airline stewardess; checking Dun and Bradstreet; overcoming the fortune hunter image; wooing the rich; and marriage contracts.

Support Systems 232
3 hours 3 credits

The value of a suitable environment which reflects the taste and position of the student is discussed. Class tours of private homes, lofts, and Bloomingdale's rooms are required along with a subscription to *Architectural Digest*. Prerequisite: "Looking Good" series.

Support Systems 302
3 hours 3 credits

A natural progression through #223, #224, #232, culminates in this class devoted to entertaining producers and network officials. Frequently termed the "Who You Know" course, ways are explored to entice useful people to your home, obligate them, impress them, and follow up the contracts. Also covered are suitable presents, called bribes.

Support Systems 303
3 hours 3 credits

As an extension of #302, the student is taught the importance of the barter of favor system of operations. Those in a position to employ others are often won over by personal favors. Covered are: designing lighting for a party, a bar mitzvah, a wedding, a funeral.

Ivied Walls 125
3 hours 3 credits

The basic course for those seeking academic careers should be taken in the same term as #116. This class takes the place of Coping 206 and stresses the first impression made by an academic job applicant. Attention is paid to tweed jackets, length of hair, beard maintenance, and pipe smoking for non-smokers.

Ivied Walls 230
4 hours 6 credits

"Stepping Over the Bodies" is the popular name for this course, which offers information about promotion and tenure procedures. The uses of publication are stressed. Methods are explored by which the student can write articles which sound impressive without saying anything. It is shown why articles in *The Drama Review* are most likely to win promotion while articles in *Lighting Dimensions* lead to demotion.

Ivied Walls 231
3 hours 3 credits

Professional work while teaching is discussed in this important class. While some faculties deplore any activity that makes money, others are very impressed by the glamour of professional experience. The student is taught to arrange his teaching schedule so that he can travel around the country on professional jobs and hardly ever meet a class. The advanced work in this course explores ways to use students as unpaid assistants.

Ivied Walls 340
2 hours 3 credits

To complete the academic series, this course proposes ways to recruit students. Areas covered are: jokes and witty anecdotes for the classroom, giving higher grades than other professors, X-rated films for classroom use.

Advanced Technical Problems 425
12 hours 12 credits

Affectionately known as "Making Mountains out of Molehills", this course covers those difficult lighting tasks which are not dealt with in most lighting courses. Ways are examined to best light aging stars, dogs, and sea lions. Special attention is paid to Restoration cleavages. Because new styles of theatre have made new demands, lighting techniques are proposed for increasing the size, three dimensionally, and importance of pudenda.

~

Application forms and further information are available from the University of Southern Forty-Second Street, Box 0, Times Square Station, New York, Iowa 470978652.

"Education At Last"!!

PROFESSION

[f. L. *professo*]

3. A vocation, a calling, *esp.* one requiring ad-vanced knowledge or training...

●・・・・・・・・・・・・・・・・・

WORKING FOR THE BOSS

First published in *Lighting Dimensions*: November/December 1985.

The call came to Vladaslaw Zxnergbniski while he was looking at a pirated tape of out-takes from *Bloopers by the Rich and Famous*. The scuttlebutt had it that a remarkable use of daylight filters created a magical moment in a short scene in which Magdelina Dinero inadvertently dropped her twenty carat diamond ring down the Crush-it-all disposal in her four million dollar kitchen. Zxnergbniski searched for anything that added to his repertoire of lighting effects.

The operator had to spell his name when he got the call because, like most Americans, she had trouble with Italian names. The people at Metro Century Fox simply called him "Slaw," which was either taken from the end of his first name or came from the salad he ate every day at the studio commissary. His name had caused him particular difficulty on the phone when identifying himself to secretaries or ordering boat shoes from Land's End. Although he had received forty-three Academy Award nominations, he had never won an Oscar because, according to his agent, no one dared attempt to spell his name on the ballot.

The operator said, "Is this Blavislar uh, uh Zee Ex En Ee Ar Gee Bee En Eye Es Kay Eye? I have a call from the White House "

"Hey, Slaw. This is Tony Rambo. Remember me? I was the third assistant director on *The Edge of Morning.*"

Slaw could hardly remember the film, much less the blur of faces that passed by while he was working. "Sure, Tony Rambo. I sure do remember. What are you doing at the White House?"

"I'm normally in the press office but I'm really attached to the image people at O.A.M. Before you ask, it's the Office of Appearance Modification. We have lots of people from the coast back here. Sometimes staff meetings look like they could take place at a Malibu bar, except that everyone is wearing coats and ties. Your name came up at a meeting this morning—it came up with some difficulty, of course. The chief, that's Bert Burt...do you remember his work on

The Russians Meet Miss America? Well, he asked me to give you a jingle today. You see, we've got a little glitch with our lighting setup and we think you're the perfect man to help us out of a snag. We checked the computers and know that you don't begin shooting on *Andy Hardy Loses the Farm* for two weeks. Could you be here at 10:00 tomorrow? It would put a lot of plusses in a lot of columns."

The following morning, feeling leaden from the red-eye flight, Slaw arrived at the White House eager to find out what important things they wanted him to do. The guards took his light meter, pipe tamper, and belt. After comparing his fingerprints with their files, they sent him off to the O.A.M.

Rambo's secretary, a gorgeous black lady named Olga with an English accent, led Slaw from the reception area to the O.A.M. wing where every wall was covered with photographs. Slaw noticed the president in every picture. "Fair enough," he thought.

Rambo was waiting at his office door. Except for the tan, he could have been an insurance salesman from Topeka. Slaw had no memory of ever seeing the man before.

"Hey Slaw, glad you made it. Sorry we don't have time to chat, but our meeting is about to blast off." He propelled Slaw down the hall to a conference room that looked like the exhibitors' area of a media convention. Instead of a conference table, the room was set up like a theatre, all seats facing a screen.

Rambo introduced Slaw to the thirteen men assembled, all of whom had connections with the entertainment industry, most as writers or producers of B movies or made-for-TV films. Sam Bush, producer of *Slave Girls of the Moon,* started the meeting with a prayer, thanking the Deity for giving those present the opportunity of serving His cause on earth and asking Him to continue guiding them in their work. Slaw realized that, given the background of these men, connections were important.

Bert Burt turned to Slaw. "Let's get right to the point. Take a look at our problem." The lights dimmed and pictures flashed on the screen.

For twenty minutes Slaw watched clips of the president's news conferences. Halfway through, one person in the dark applauded. Rambo whispered to Slaw, "Smithers has done only one thing while he's been here, switch the president's podium to the front of that hallway. See how the chandelier makes a halo over the chief's head? Smithers won't let us forget that."

Apart from very excellent lighting that made the president look younger and bad lighting that made the press look older, Slaw noticed that the president was persuasive while reading a text but stumbled around like a drunk turkey while answering questions.

The lights went up. Bert Burt resumed, "As you saw, the president is persuasive when reading a text and like a dru... I mean, not so great when answering questions. Our assignment is to remedy the situation. It is of prime importance to the safety of the free world that we present our leader as an in-charge man whose control of every situation is absolute. Because you are the best lighting director in the world, we're asking for your help."

Everyone watched Slaw. "Don't let him answer questions," Slaw ventured.

"Not possible," Rambo said. "Our contract with the press stipulates that they will get an agreed upon number of on-camera exposures in exchange for favorable treatment of the president. The question period fulfills that obligation. Besides, answering questions has become a tradition; its absence would be noticed."

Bert Burt took over; "We have thought about most everything. Two possibilities have weathered our deliberations. Both are technological, demand the utmost in sophisticated use of lighting, and are perfectly possible within a $423 million budget. We call the first possibility 'Maximized Response Pre-inquiry,' or M.R.P. for short. It works like this: reporters, in exchange for their silence on the method, will be allowed to tape a question. If they flub, they can do as many takes as they need. It makes them look good. We take the tapes, prepare answers for the president, and he makes a tape. We splice the whole thing together for a technically perfect live news conference."

Slaw was ahead of them. "I see what you need. The lighting has to be matched so that each tape looks like it was done in the same room at the same time and under the same lighting. We do that all the time. Easy. What's the second scheme?"

"The other is called C.E.E.," Burt replied, or "Computer Enhanced Elucidation." As you know, it is possible to use the computer to index just about everything. In this case we make a file of every possible question that might be asked at a press conference and pair those questions with pre-recorded responses (using, of course, words and phrases already spoken by the president so that he is not exhausted from recording hours of text). At the press conference we need only punch in a reporter's question to access the proper answer."

Slaw didn't understand. "I don't under stand," he said. "The president will just stand there while his voice comes out of a computer?"

"He will lip-sync." Burt smiled triumphantly, pleased with the boldness of the plan. "Can you light him so that any gaps are covered?"

"That will take some doing...but if the angle is right and if the director chooses shots fast enough and if the president is skillful in his lip movements...Well, there just might be a chance."

Suddenly the room was filled with men shaking hands and patting one another on the back. Slaw heard the phrases, "We did it," "Great," "Bravo Us," "Thank You Lord." Burt called everyone back to order, "Gentlemen, gentlemen! It looks like we have two options for the boss. I suggest that I take Mr. Ghnen...uh, Slaw upstairs right away so that he can tell the boss himself."

"Do you mean I'll get to meet the president?"

"Oh, no. We're going to the first lady."

POWER, GREED, SELF AND HUMILIATION

First published in *Lighting Dimensions*: March, April/May, June 1983.

PART I

On a stool in the corner was the designer's assistant Bob Cratcheting. Between drafting and sharpening pencils on a sandpaper block the young man looked up to catch the conversation on the other side of the room.

I was interviewing the designer. After forty years of work he still remembered the slights and wounds of his early career. He was telling me about his first Broadway musical.

He recalled showing the sketches for a particular scene to the director for approval. The director was ecstatic. The sketches were the best thing he'd seen in theatre since Ethel Barrymore showed an ankle. Words could not describe the excellence of the designer's work. The director was so pleased that he suggested they rush over to show the producer these wonders of theatrical skill.

The playwright was just leaving when they arrived at the producer's office. He took one look at the sketches and said that the set was not at all what was needed or intended for the scene. Turning to the director he said, "What do you think?"

Without a pause the director replied, "I told the designer exactly that. This set is not right."

Now, many years later, the treachery of that moment remained in the designer's memory as an example of immorality in the theatre.

That evening I got a call from the designer's assistant. He told me that he could not help hearing our conversation and had to add to what had been said.

"Recently, I was in charge of a setup, representing the designer, and trying to make everything go smoothly. Just before lunch break there was a dispute between two crews about which jobs should be done first. I could not settle it because it depended on whether or not some extra lighting equipment would be hung. If it was, work had to stop on everything else.

"I called the designer immediately to report the difficulty and he thanked me for letting him know so promptly. He called the manager and they agreed to meet at two o'clock on the set. The problem was easily settled, but the manager gave me hell for not reporting it sooner. Saying I had wasted crew time, he called me incompetent and some other choice things that I don't remember.

"I expected my boss, the designer, to defend me. After all, two hours earlier he had thanked me for my efficiency. Instead, he turned on me and agreed that it was my fault for not letting them know earlier.

"I thought that you would find it interesting that although my boss remembers the wrong done to him by the director years ago, he is capable of exactly the same maneuver to make himself look good.

"What makes both cases so pitiful is that they were such petty matters. I thought that once one achieved some success, it would be possible to accept responsibility for things going wrong. That doesn't seem to be the way it is. No one is ever big enough to accept blame, even for little things. Everyone spends time covering his own ass. I wonder if when I have an assistant I'll dump on him to make myself look better. This is not the sort of thing I wanted to learn as an assistant."

The following article developed out of many hours of interviews with theatre people like the designer above. I talked to designers, directors, actors, dancers, and a psychiatrist. I have not used their names here because the information they gave me was often damaging to others. I hope that those who gave me help and were candid in their comments know that I am deeply appreciative without publicly naming each of their contributions. Those quotations that are attributed can be found in published work. The other quotations are sometimes edited for length and to disguise names.

When I asked theatre people about ethics in theatre, they almost invariably said there are no theatre ethics. But as they recounted tales of terrible things that were done to them or to others, it became clear that they were measuring the acts they described against some standard of correct behavior. This seemed entirely reasonable to me because no business, craft, or art that depends on the collaboration of so many people can exist without some rules, even unspoken ones. The most basic primitive tribe has some basic taboos. As theatre developed over the centuries, it is logical to expect that rules evolved to govern the relationships between individuals and between the group and individuals. As just one example, union contracts indicate that disputes have occurred in certain areas, that they have been settled by negotiation, and that anyone breaking these contracts departs from a moral norm.

I became interested in theatre ethics several years ago when a set I had designed was rented by a regional opera company whose director thought that the painstakingly aged walls of my set were too grim for his Sunbelt audience. He had the set painted bright, flat, pastel colors. He did this without permission and against provisions of the union contract. (When the set was returned and about to be used again, it had to be repainted.) I was outraged. Did this director have no sense of professional ethics?

"...no business, craft, or art that depends on the collaboration of so many people can exist without some rules, even unspoken ones."

As I spoke to others concerning their concepts of ethics and morality in theatre, I realized that this subject is monumental. To be properly studied, it requires a polymathic understanding and years of full-time labor, neither of which I have. What I have attempted to do here is to set down some random impressions under categories such as money matters, working conditions, personal relationships, understood obligations, historical factors, and philosophical and religious influences, along with my own very subjective opinions. I have certainly not tried to formulate a code of conduct. My search was to better understand why we act as we do and to see what we expect from others in common situations.

Since ancient times, theatre people have been excluded from society. Now and then a darling of the crowd or a genius has been grudgingly accepted. But for the most part, theatre meant mountebanks, thieves, whores, a group to be avoided. The stereotype of the immoral actor was probably based on some truth. Forced into a gypsy life, the actor felt no weight of continuing local relationships to keep him moral. Nevertheless, the traveling companies must have had their own codes peculiar to themselves, concerning such issues as the division of money and what plays to perform.

Part of the theatre code was the well-known rule, "The show must go on," which led theatre people to lie, cheat, and steal. This admirable internal precept, which fostered an extreme pragmatism, led to actions deemed immoral in the wider society. War and elections are possibly more pragmatic than theatre, but short of these it is hard to imagine an activity less guided by moral principle, or one in which deadlines, and ultimately public acceptance, combine to convince the participants that commonly held moral considerations should be ignored in order to reach a goal.

Contributing to this pragmatic ethic is the pressure that binds one to a group and simultaneously separates the group from the rest of the world. The binding power is the dependence each member of the group has on all the other members. In theatre, each performance depends on cues being given, props being in place, costumes ready, every part of the entire enterprise progressing with precise timing. The whole depends on how each member carries out his assigned tasks, no matter what difficulties stand in the way. Somewhat like an athletic team, the theatre troupe concentrates on winning, but has no game rules to guard against fouls.

The team spirit operates more in educational theatre than in professional theatre. Looking for props or costumes, it is not uncommon to find students who "borrow" furniture and clothes while no one is watching. I remember a student in New York who showed up with a pay telephone for a set, a phone that Ma Bell thought was still collecting dimes on some subway platform. Students who would not think of stealing for their own gain (a generous presumption) do so for the good of the group. Under the pressures of being relied on, of an approaching opening night, people substitute one moral standard for another.

In professional theatre, while reliance on others is still necessary, the good of the group is often at variance with the ambition or the ego of the individual. In these situations, which we will look at later, ethical questions become very real as the good of the individual versus the good of the group changes from an abstract problem into a constant dilemma.

On a philosophical level, Socrates prepared the way and Plato developed the basic concepts of ethics which many people adhere to: good, doing the right thing, depends on the knowledge of good. No man voluntarily chooses to do the wrong thing. Lack of knowledge and truth is the cause of unethical behavior.

According to Plato, theatre people, more than all others, seem less likely to grasp truth. They are notorious for living in an unreal world, peopled by imaginary characters, changing persona from one production to the next, and simulating emotions. They are therefore less able to behave morally. Plato warned his students that actors, portraying the evil as well as the good, the weak as well as the courageous, were not to be trusted.

Throughout history, philosophy and religion have warned against the immorality of theatre. The Utilitarians were upset because theatre did not point the way to a better society through the redistribution of wealth. Nietzsche started out supporting great drama, especially Wagner, but ended by turning against his former idol. Certain critics today object because theatre does not reflect their own ideas of what should be taught or what should be shown to the public.

Although philosophical criticism remains important, nowadays we tend to look at ethics in more psychological terms, stressing emotional foundations of behavior over what has in the past been an external set of rules imposed by fiat or revelation. We might say that the external rules are used as ideals to be understood intellectually while we struggle with emotional forces that pull us away from the ideal.

In light of this, the work of Geoffrey C. Hazard of Yale is interesting. Writing about professional ethics, he says that ethical decisions come down to choices that affect our egos, our concepts of self, our own identities. Identity has always been a problem in theatre. In other professions, one's self image is usually tied to the opinion of one's colleagues or the professional community at large. In theatre, however, the ego is often, though not always, also tied to the opinion of a large general audience. Sometimes the adulation and respect of an audience can be won at the expense of the good opinion of one's colleagues. Therefore, making an ethical decision in theatre can be a choice between two kinds of ego gratification. Consider, for example, the star who has a supporting player fired because he or she is too good, or the director I interviewed who told me that his job was to get the best performance possible, even if it meant doing emotional, physical, or professional damage to an actor.

There were many contradictions and conflicting forces that people in the theatre told me about: struggles between directors and designers, between directors and actors, and between everyone working on a production and producers. But the first thing that everyone talks about when discussing ethics has nothing to do with the psychological or emotional stresses that make people behave as they do.

The first thing professionals talk about when asked about ethics is financial morality, or the lack of it. Being cheated out of money is more easily seen than is a surreptitious attack on ego. Stealing money can take the form of kickbacks, usurious contracts, or nonpayment for work done. It also includes the theft of ideas that produce money. What separates theft in theatre from the same shady practices in the business world is that the equation in a theatre deal pits the artist, a non-business person, on one side and a manager producer, a business person, on the other.

Hal Prince sees both sides: "Commitment to theatre is a wonderful thing, but it costs. We love working, and we are frustrated by those business elements which we must respect in order to keep working creatively. Union regulations frustrate us because we ask and I think with some justification, how can you regulate creative activity? The trouble is that some amount of regulation is necessary, that artists are taken advantage of because they are artists that entrepreneurs can be sons of bitches."

I suspect that there are two different problems that Prince is writing about. One is the union worker who is not engaged in "creative activity," but is expected to be completely committed to a project even though he could be, and often is, replaced by another "body." As a replaceable cog in the machine, the non-artist worker will get as much money as possible and will do his work well, though not with selfless dedication. I was told that in a regional or repertory company, where the good of the whole group is at stake and workers are employed full time, there is a different attitude. One designer told me that he preferred to work regional theatres for just this reason. He told me that away from Broadway, the work is its own reward. No one thinks much about money, because there is none to be made. Others feel that regional, off-Broadway, and Broadway work is all the same, except that you earn more on Broadway. Almost everyone told me that as the money to be made grows larger, so does the greed. As this happens, management, or as Prince says, "entrepreneurs," attempt to squeeze every last penny out of a production.

"I have never done a show in which I have not been screwed by the contract."

Management is able to squeeze more from the creative people because artists are led by two carrots on two sticks: one feeds the ego, the other provides real food, clothing, shelter, and an occasional glass of wine. Manipulation of the artist is easy because the artist's ego is often more hungry than his stomach.

The standard management arguments are, "This will be great for your career; all the critics will be there; you'll be discovered; it will look great on your resume." Although these inducements are seen as fictitious by both management and the artist, the artist accepts them because above all else he wants to work at his art. (Later we will see how ego and a need to create are combined.)

In negotiations between management and artists, another powerful force gives management an edge. So many others are waiting to get one job that ninety-five percent of those I talked with cannot say no to an offer. Very few artists can afford to name their price and say take it or leave it. A whole lot of shows are done for minimum wages.

Union contracts regulating the relationship between artist and management are basic and uncomplicated. They stipulate: work to be done by the artist; time in attendance for rehearsals and performance; pay received by the artist; pay for additional use of the artist's work; repayment of expenses; artist's rights when his work is changed; credit given in programs and ads; procedures for firing an artist; arbitration of disputes; and prohibition of discrimination.

The existence of these basic rules implies that over the years, the same sort of disputes occurred, that artists were taken advantage of, and, to a lesser extent, artists did not fulfill their obligations. But because the contracts are basic, several of those I interviewed complained about the contract procedure.

"I have never done a show in which I have not been screwed by the contract. I sign a few contracts a year and the managers deal with ten a day. They know what they are doing. They are the wolves. I am a sheep. After each unfortunate experience I added an extra clause on the next contract, but then I was screwed in a new way."

"One designer designed sets for a Broadway show, but found out that the out of town theatre made the Broadway sets impossible. He ended up designing two shows for the price of one. You agree to do a job and you do it no matter how long it takes."

"The union contracts deal only with business but never care about the quality of work. Something about the quality of work should be included, but they shout you down at the union meetings."

"I had to give up getting angry or upset when I'm screwed. I would be angry all the time. It would ruin my life. I couldn't think of anything else."

From the stories I heard, there seem to be two reasons why artists get cheated on contracts. First, in any uncertain situation, management has a better chance of having protected itself in the contract. Second, the artist's "work" is the production, while the manager's "work" is the contract. The successful contract for management takes something away that might have been granted. In some ways the managers enjoy playing a game in which there is a competition, where someone wins and someone loses. The manager's ego demands the game.

Although it is not related to theatre, a book was published in 1977 by Financial Management Associates Inc., titled, *Why Sons of Bitches Succeed and Nice Guys Fail in a Small Business*. The flyer tells us, "In Chapter Two we will show you how to screw your employees first (before they screw you)—how to keep them smiling on low pay—how to maneuver them into low pay jobs they are afraid to walk away from—how to hire and fire so you always make money." If this book speaks to even a small need in the businessman, including theatrical managers, it is clear how wide the gap is between artist and management. One story told to me was about a famous producer who likes to play games with his employees. Equity ruled that an actor could understudy only four roles. The producer had four shows running and "made one poor son of a bitch run back and forth between four theatres at show time."

Besides competition and game playing, there exists, of course, real naked greed. As a young art director on a film, I bought one dollar's worth of nails or tacks to put up something in the drafting room. I conscientiously gave the outside-prop man the bill and watched him add an extra one in front of the total. When I questioned him, he shrugged off my concern: "That's nothing, the production manager will add at least another digit to the left of mine." My one dollar purchase grew to two or maybe many hundreds of dollars on the books.

From one designer: "There are deals between the manager and the prop man. There are always deals between the manager and the prop man."

Another designer: "I designed chandeliers. The manufacturer sent back a price that was double what I expected, so I went to see him. He told me the electrician had told him to double the price for his kick back. The producer fired the electrician on the spot, but the director got him back. He was his right-hand man."

Another such tale: "Working for Rogers and Hammerstein, I found that the electrician and the manager were crooks. I told Rogers who said, "Yes, we know. But he has to deal with so many crooks we think he is good for the job."

Stories like these go on and on. To be fair, I could cite other stories of generosity and honorable behavior, but I heard these less often than the tales of greed. I suspect that ethical acts are actually more common, but because they are the rule, the exceptions are more easily remembered and recounted. Furthermore, victims carry scars that are ready reminders of the wounds they suffered.

Except for the weak position of artists in relation to business people, honesty and financial ethics in theatre appear not unlike the rest of society where theft, kickbacks, bribes, influence, and pressure are part of the landscape. Making theatre people more honest would require making the world more honest. As Elmer Rice once wrote about the evil effects of money and commercialism on theatre, "…in so far as the evil does exist, it is an integral part of our society, and could be eradicated only by drastic alteration of our psychology, cultural habits, and economic institutions."

Necessary as money is to all of us, the most damaging injuries suffered by those I spoke to were not caused by being cheated out of money.

"The lack of ethics on the part of a producer has to do with greed. The lack of ethics on the part of a director has to do with ego."

"I prefer a genuine crook whose motives you understand. Then you can protect yourself. I prefer him to…, who was an intellectual gangster who talked only Shakespeare and art, but was really in the back room trying to cut your heart out, undermining you and destroying you."

PART II

In matters of money, there are historical, literary, philosophical, and religious precedents against stealing and cheating one's neighbors. But the codes dealing with human relationships are less clear. Often parties in a relationship cannot agree even that an immoral act has been committed. As an example, confusion arises when what a director does to an actor is seen by the director as moral but by the actor as an unwarranted attack on his ego.

One can choose any number of moral precepts by which to judge an action: "Do unto others as you would have them do unto you." Or Kant's: "Act as if the maxim of our action were to become by our will a universal law of nature." Or: "The greatest good for the greatest number." We can look to religious law or to new theories of moral behavior by such people as Lawrence Kohlberg. It is hard to know what to think. As example, by the time Kohlberg and his six steps of moral development are influencing teachers, he is attacked by the feminists who criticize his theories for being based on boys rather than girls and assert that each sex has different moral imperatives. Contemporary educators tell us that the family's power to influence a child's conscience is giving way to the pressure of his peer group. Now a child's super ego may not be based on older, more universal, concepts, but on learning only the rules for the "cooperation game" within a very small peer society.

The theatre professional is also a product of family, place, religion, and/or a greater society, all of which have combined to make his super ego different from that of his colleagues. At the same time, theatre, as a small society with its peer group pressures, creates rules that affect this small society as a whole. What I found in talking to people was not a clear set of rules to clarify behavior in the theatre, but rather certain recurrent ideas that have one foot in theatre ethics and the other in general ethics.

One of the most constant phrases that cropped up in my interviews formed a pattern. "When you get up there in the jungle you have to protect yourself."

"Survival has to do with strength and luck. It's a jungle."

"Your ass is on the line every step of the way. Everyone is protecting himself, protecting his own turf."

The image of the jungle, with wild animals hunting and being hunted, was repeated many times. The territory is treacherous. A misstep spells death. If another is weak or wounded, pounce! Yet, if one accepts this image as a realistic picture of human relations in theatre, one misses underlying implications. For one thing, the more someone rages at jungle morality, the more he demonstrates that an assumed code is being violated. Somewhere, either from earlier family training or peer examples, or from divine instruction, by recognizing that theatre is a jungle, one also gives testimony that there is a higher morality, a better way of social interaction, an ideal of human behavior that is being violated.

There is, furthermore, a hidden set of emotional reasons for taking part in the jungle life. As Dr. Donald Kaplan puts it, "The feeling of being victimized is too often only the conscious aspect of an active but unconscious negotiation with the victimizer we complain about. Such a feeling should not be taken too quickly at face value."

We will look at the trade-offs for being victimized later, but for now we have to muddy the waters further by adding another theme: Several people see lack of communication as a major cause of misunderstandings that lead to unethical behavior. A lighting designer told me that people who can't communicate have no chance in theatre these days. "They must understand people." This idea has led to a renewed interest in theatre people working together. Robert Benedetti, for one, calls for everyone to be aligned toward the same goal: to make the play a unified event. He asks for trust, support, respect, and open communication.

If there is now a need for trust, support, respect, and open communication, they are not present in very many theatre relationships—and they are exactly the ways of behaving that are moral. Without them there is only the jungle. So let us return to the jungle metaphor. It is as if lions, eagles, snails, and elephants were trying to produce a "unified event." Each of the species has its own goals, interpretation of reality, language, and unconscious negotiations. If disagreements arise from misunderstanding, even with the greatest goodwill among parties, the defense mechanisms of dissimilar endangered jungle creatures make cooperation unlikely. How can you be open with someone who will gobble you up at the first opportunity? Ethics may be the luxury of the secure.

I don't want to go on and on about animals, although they seem to be a common way of discussing theatre, but the next step in looking at theatre relationships must be the pecking order that exists among the variety of tasks to be done. By custom, the director has the greatest amount of authority. He is a father figure (or mother figure, if you will). Several designers told me that one rule that governs their ethical life is that in a production, the director's vision supersedes everyone else's. An unwritten code demands that everyone do what the director wants. Designers and actors might be surprised to hear that directors feel very vulnerable.

Hal Prince writes that in *Superman*, an actor whose lines were being cut to shorten the play invariably pulled out a scrap of paper and in front of the whole cast suggested a new line to replace one of his cut lines. His suggestions were rejected each time but this did not stop him. "I realize how castrating an actor can be if he chooses. We are all familiar with the occasions in which a director can bully an actor into confusion, but there are as many times when actors run off with the play. Stars do it all the time."

Directors feel victimized when their authority is subverted and when they are unable to get others to do what they wish, even though an honest effort is being made. In both cases the result is tremendous frustration. And as a director told me, "The pressures of frustration make creative people cruel."

If frustration and cruelty are intertwined, the director must be most cruel when he is most frustrated by the inability of an actor or a designer to do what is wanted. "The perfection they see in their imaginations can never be satisfied, so they feel you are killing their baby," was the way one man put it. Or as another saw it, "The worst agony in the world is having your artistic vision trampled on."

A director described a scene in which an actress has tried twenty times to do a speech the way a director wanted it done. Does the director ask her sweetly to do it again, or does he scream, insult her, and throw a fit? The man I talked to asked, "Are we going to treat people like human beings or are we going to get the scene right?" He agreed that the polite way was the moral way, but added, "A great artist is not a normal person."

And so we add another dilemma into our considerations: Do we excuse great artists their moral transgressions but damn the majority of ordinary artists who have the same or greater frustrations? Do we, by accepting the great artist's immoral behavior, encourage the lesser artist to behave the same way in order to seem great? And who decides who is, or is not, great? The question is one of condoning immoral treatment of others by anyone, great artist or not. Is there a line that can be drawn by anyone between "acceptable" and "unacceptable" immorality?

"One scene designer, knowing he will not get all the scenery he wants, designs more than is necessary, so that when compromise and cutting occur he ends up with exactly the sets he really wanted."

One of our greatest directors has a reputation for dreadful treatment of others. Certain artists will not work with him although his work is always exciting and often brilliant. For them the risk of psychic damage is not worth it. One story about this man involves his abuse of a lighting designer. For hours the designer was told he was worthless, incompetent, and worse. Each dimmer was taken up a point, down a point, usually to be left where it began. The director pushed, insulted, swore, and badgered for a whole day. At the end, he turned to the designer, and said, "You know, it has taken eight years of analysis for me to treat you the way I have today."

Designers react to this story by shaking their heads from side to side. Sympathetic pain clouds their faces. Directors react by nodding their heads up and down. "Yes," one said, "I understand the insecurity, the desire for perfection that makes a man act that way."

Moral questions are matters of choice. In theatre there appears to be a choice, at times, between people and product, just as in the larger world there might be a choice between people and property or in a holy war between principles and lives. However, I suspect that much of the bad behavior we see in theatre is not considered, is not a moral choice, but is an emotional reaction to frustration. Such reactions are childlike. As we grow up we are trained to control our frustrations, suppress our anger, use logic and reason to get our way. And it might just be that the director, the one who in theory has the most control over others, is also the one with the least control over tangible, pliable, manageable elements of a production. It is he who must exercise the most adult forms of controls: logic, reason, persuasion. If these fail he has no other tools. He is frustrated. "Frustration makes artistic people cruel."

A designer works more with objects than with people. He has material things to manipulate and bring under control. Much of his artistic vision has been worked out at the drawing table. If he is not getting what he wants this may be largely his own fault. The designer is faced with ethical problems when his artistic vision is in conflict with the artistic vision of others, mostly directors and producers. The way designers handle these conflicts was a major topic in my interviews, and designers were not a bit hesitant in telling me of the jungle wiles they use to get their own way.

One scene designer, knowing he will not get all the scenery he wants, designs more than is necessary, so that when compromise and cutting occur he ends up with exactly the sets he really wanted. This can backfire, as with the designer who was told not to cut and had to use more scenery than he had intended.

Lighting designers have their own deceptive ploys. Several told me that during the rush of rehearsals there is no time to educate a director about visual perception, so they absolutely must use a few tricks. For example, when a director wants more light on a performer but the lighting designer has no more light to give, or feels that there is enough light already, he may appear to agree to the director's request. He takes down all the lights except those on the performer. There are times when a director asks for more light and the designer knows that he will repeat the request. The designer may then have a code with the electrician; with the director at his elbow, the designer calls for certain circuits to be brought up to seven; the electrician takes them to five. The next time the request is made to bring them to nine, they may go to seven, and so on. In most cases directors are perfectly satisfied and deceived.

When asked about the morality of fooling a director, designers contend that the director is really searching for any change because he doesn't know how to correct his own direction. He fixes on the physical production because changing a dimmer reading is easier than changing an actress's line reading. It is not splitting hairs, however, to question the use of any deception. Sissela Bok in her book *Lying, Moral Choice in Public and Private Life,* stresses the immorality of any lie, white, black or gray, because there is no logical point at which one stops lying. Shakespeare's way of putting it was, "a quicksand of deceit." In terms of the designer and director, when does a designer stop tricking the director? Which director does he trick and which not?

There are ways a designer can honestly try to get a director to modify his requests. One lighting designer told me that he gives a director exactly what he asks for, and this is often so awful the designer wins his point, assuming that the director is still able, in the confusion and panic of final rehearsals, to tell good from bad. Another designer said that he asks to show his way first so that the director can see how good it is. If the director still wants to change the lights, the designer obliges. He went on to admit, sheepishly, that when the director insists on his way and all else fails to convince him that the lights are wrong, the designer may, over several days, gradually bring the lights back to what he wanted originally.

This same designer rails against the unethical attitude of managers who tell electricians not to check the lamps each day. The electrician, he says, is loyal to either the manager or the designer; if it is the manager, there are likely to be burned out lamps at each performance, cheating both the audience and the lighting designer.

Several designers spoke of the immorality of cheating an audience, or, as one put it, "the ethical problem of how to treat a work of art." I was particularly sensitive to this after an opera set I designed continued to be used all over the country although it was in shambles. Legs were missing, pieces were torn, drops were dirty, and a scrim was gone. Each time it was used the audience was given less than they had paid for and my reputation suffered.

Reputation, artistic integrity, and ego are tangled in a confused but nevertheless powerful force motivating designers. They want to see their own visions realized on stage. The young designer, especially, is often ruled by the emotional power of ego because his ego is more fragile and less protected by an armor of past success. The older designer can tolerate his vision being questioned and can accept the possibility of other ideas and other visions. At times, however, both young and old designers are faced with demands they feel are wrong. To acquiesce would damage the show and their own reputations. Their only choice is to quit the show.

With no objective proof whatsoever, I think that the number of designers quitting a show is a thousand times greater among young designers, in small, low visibility productions than it is among older designers on large professional productions. Possibly the older designer has learned to compromise. Maybe, as with university professors, there is more fighting when the rewards are low. The real reason, I believe, is that the kind of ego satisfaction supplied by the small production is in a large production complicated by many other layers of conflicting needs. The balance between artistic morality versus artistic whoring shifts as the stakes grow higher in money, recognition, and power. Some people feel that the artistic impulses that have brought them to theatre have been overshadowed by the pressures of playing for high stakes. To quote the choreographer Paul Taylor, "As you become more and more successful, you become more and more like real estate, and that's not what art is about. Art is about taking risks. Danger and chaos—those are the real muses an artist must court."

To quote some of those interviewed:

A designer: "A producer told me, 'I've hired you, paid you so much, you're mine.'"

A director: "If you are needed for the success of a project you are treated like a king, if you are not, like a shit."

A director: " There are very few artists in high places. It is all hit or miss in choosing people, the choice of plays, actors, designers. If you have a hit, everyone wants you to do everything—not just the things you can do. It is all mixed up with success. You are as good as your last show."

At the highest levels, where there are great rewards, there are also confusing and contradictory pressures that cause people to behave in ways amazing even to themselves. One director said that after his first big success, when he was sought after by everyone in town, he locked himself in his room and drank, feeling inadequate and terribly frightened. Others can enjoy success, even if it lasts but a short time. But when actors become stars, they often seem to go through personality changes that alter their relationships with others. It is difficult to maintain an equilibrium when normal feelings of inadequacy, which have been reinforced along the way by miserable treatment, are contradicted by an external world that suddenly showers one with money, attention, and deference. Values, relationships, and the perception of reality can shift drastically.

Depending on your point of view, designers are protected or denied the excesses that come with stardom. Designers are never safely above abusive treatment. They are never paid astronomical sums of money, recognized on the street, or chosen to be the subjects of film biographies. One of the greatest designers of our era, at the end of his very distinguished career, was fired and cursed in front of the entire company of a play. This brings up an interesting ethical practice adopted by designers. If one designer is fired from a Broadway show, the next one hired is expected to check with his predecessor before accepting the job. In the instance above, the new designer, due to a mix up, did not talk to the other designer and was criticized by his colleagues.

Contrast the way designers treat each other to the following recollection. A director told me that in the old days, early in the rehearsal process, the Theatre Guild used to have extra unemployed actors sitting like vultures in the balcony, watching. If one of the actors was fired, there was always a ready replacement who knew the blocking. This same director jokes that to survive, you must learn how to recognize your replacement in the hotel lobby out of town.

There are several reasons why actors and designers treat their colleagues differently. First of all, there are far more actors than designers. The small number of designers makes it possible for them to know each other. A lighting designer told me that his colleagues see each other's shows, discuss their problems and, in effect, have a fraternal feeling. This is not to say that designers are always sweet and lovely to each other, only that it is harder to do nasty things to someone one knows than to a stranger. By contrast, actors work with other actors in situations that are competitive: who gets the best notices, who gets the biggest laugh or applause. Added to the competition is the possibility of lightning striking, making one actor a star while those around him struggle on. Another factor is that designers can step back and look at their finished work in a way that actors never can. Given the opportunity for an objective evaluation of their work and their image of self worth, the designer may not be as insecure as the actor. Insecurity and competition contribute to the possibility of harmful behavior.

Insecurity is present even at the top where stars are not only still competing but are at the same time given the power, indeed encouraged to exercise that power, to behave towards others in ways that are not normally acceptable. While many stars are noted for their kindness and "one of the gang" attitude, the very fact that a crew or the press labels a star "easy to work with" indicates that the nice guy at the top is a rarity. And even he or she is suspect. A director observed, "If you think someone has survived a lifetime in theatre behaving honorably and morally, you just don't know their story."

There are maxims about success in the larger society that mirror what I heard about theatre. Success is said to be achieved by stepping over the bodies of the vanquished. It is said that power corrupts and that those at the top have different moral imperatives. All this may be true to some extent, but I think there are differences between success in theatre and success in the civilian world.

One remarkable difference is that theatre produces a divisible product. Critics can and do separate writing, acting, directing, and design in their reviews. Imagine a story about a new product that not only singles out parts of the product for comment, but also names those responsible for design and construction. "The welding of the bumper of the new Ford pick-up by Henry Talbot was very smooth. The fitting of the plastic dash by Mary Palmer, however, was second-rate and did not accomplish the intentions of Joe

Parker, the dash designer." Absurd as this example is, it points up two differences in theatre life. First, the end product of theatre is still seen as a combination of individual talents. Second, the work of theatre artists is so interesting to a public that it is worthy of news stories as well as critical pieces. Little wonder then that even stars are insecure when their work is so widely noted. Little wonder that the ethics of team work are so difficult to maintain.

The stories are legion concerning actors upstaging fellow actors. And we must not forget the designers whose flashy work is critically praised but may hurt a show. These examples raise the question of what most strengthens an artist's ego—the confirmation of the public, or the respect of fellow workers? One hopes that these two methods of ego gratification are not mutually exclusive. It may be possible to be praised by the crowd and one's peers at the same time. But given an ethical choice, choosing an action that either makes oneself or the group look good, the weight of fame, power, and money lies on the side of an audience's acceptance while theatre people pay lip service to the ideal of ensemble playing.

The concept of ensemble playing, which is more honored in the breach than in fact, was only one of the contradictions that ran through the interviews I taped. One of the most confusing dichotomies was the way everyone complained about the lack of morality and therefore implied that there was a standard that they looked to themselves. Beneath the surface that was described as a jungle, there was an ethical sense that was both judgmental and accepting. One of the most cynical and critical of those I interviewed was, from my own personal relationship to him, a most caring and moral person. He told me, "Humiliation is the lifeblood of the theatre. I don't know anyone too big who can't be humiliated in theatre."

I had heard opinions like this from so many others that I had to ask why, if theatre is a jungle where everyone is cheated and humiliated, is anyone willing to submit to a life in theatre? I had to go one more step beneath the surface and see what kind of person works in theatre, what are the hidden rewards that make it all worthwhile, and why the personalities involved are in conflict with ordinary forms of morality. The next article will examine these problems.

PART III

I talked to a psychoanalyst who is an authority on artists and has treated many theatre people. He was not surprised by my question for he has long thought that there are hidden forces at work in the personalities of artists. Artists are different.

He believes that artists have a view of the world unlike the reality of ordinary people. They inhabit a dream world, a visionary world, as well as living in the world the rest of us live in. They must make the rest of us share their second world or else they are like madmen whom no one will believe. A fine line has always existed between art and madness, between the man who sees visions and the man who makes the world share his visions.

The need to make the world accept his vision is a powerful force in the personality of the artist, a force not present in the non-artist. It explains many of the problems we have seen in relationships among theatre workers. The director may have a vision of the play as a whole, a complex world that he imagines on the stage. The designers may each have a slightly different vision, while the actors probably have many different and separate ideas of how their parts, at least, should be played. In the end, all of these dreams, fantasies, and images must coalesce to satisfy all the artists involved as well as communicate this shared vision to an audience. The likelihood of all of these things happening sounds impossible; when a great work of theatre art is created one senses a miracle taking place. Most of the productions we see are in some way flawed, just as most of the artists participating in a show are not quite satisfied—or downright disappointed.

In the process of putting together a play, two sorts of compromises are necessary: artistic compromises, in which differing visions must be considered, and practical compromises, in which non-artists (managers, producers, backers) demand changes based on money. In the first case, one hopes there is enough mutual respect among the artists to change compromise into collaboration. However, when one artist has a stronger vision, and a stronger need to see it realized, in other words is closer to madness than the others, often he prevails, leaving his colleagues frustrated. We know of great artists who are impossible to work with, whose personalities are close to disordered, but yet are able to show us wonderful worlds of which we have never dreamed. Sometimes we forgive them their immoral behavior, their running roughshod over the ideas and feelings of others; other times we cannot. In a way, our forgiveness parallels the legal defense: innocent because of insanity.

Compromises hammered out with agents, managers, and producers are equally frustrating to the artist because business people do not enjoy an artistic vision and thus speak a different language from artists. Compelled to rely on business people in order to communicate their visions, artists are resentful. The psychoanalyst told me that every sort of artist funnels his work through managers, agents, editors, and gallery owners who take advantage of his need to create. "The artist, except for the recognized tops, always feels exploited."

So far, I have lumped all kinds of theatre artists together. In truth, the people I interviewed, especially the psychoanalyst, separated actors from other personality types working in theatre. In the majority of opinions, the actor is the artist who is most exploited and most humiliated. He stands on stage and suffers insults from directors, rejection at auditions, and embarrassment by fluffs in performance. He must deliver unfunny jokes and play maudlin love scenes, then be reviewed both for his abilities and his physical appeal. More than any other artist, the actor is vulnerable and exposed.

The analyst explained the actor's situation this way: "Actors have a very fluid identity. They have very little sense of self. An actor can go to a party and look from outside himself at what he is doing, as if he were observing a scene. In others this is called depersonalization and is a real madness; in actors it is usual."

One of the problems resulting from the actor's lack of identity has been labeled by a director as "bad bringing up in the theatre." He says that it is almost impossible to speak to an actor in technical terms, thus it is impossible not to wound an actor's ego. "If you tell an opera singer that the part requires a dramatic soprano and she is a lyric soprano, she will understand and not be hurt. But if you tell an actress that the part requires a dramatic personality and she is a lyric personality, she won't believe you. She will say, 'I can act anything. What do you want? Just tell me and I will do it.' The art of acting has never been defined. Everyone thinks he can do anything and he tries to do everything. Nobody is ever corrected in technical terms. The theatre has never been considered an art.

Roles are not defined because it looks like life up there. Therefore everyone thinks he can be an actor."

Obviously without technical criteria with which to evaluate his work, the actor is likely to take rejection as a personal insult. He sees the actions of a director or producer as immoral. The analyst advises, "To protect against feeling victimized, an actor can try not to take his treatment personally. Also, the actor can try to separate what is going on from what he thinks is going on."

To complicate matters, the actor's job requires submission to a director. It is well and good to speak of collaboration in theatre, but if one person has final say, there is always a degree of submission present. The analyst told me that not all submission is submission to tyranny. It only seems like tyranny if it is done with resentment. "Some directors have earned their authority and those who haven't may become authoritarian. But some actors can not respect anyone. Any submission is a loss of self esteem. These actors probably have less sense of identity—of self."

An actor working in a play travels an interesting route. At first he deals with the director, described as the paradigm of a parent. "He can be benevolent, cold, strict, sadistic. The actor resents the infantile position of submission. Thus he usually complains about being victimized." The actor looks for an ideal parent, one who may be controlling but is worthy of trust. The actor needs to feel that this substitute parent has the actor's best interests at heart.

The payoff for actors comes in performance. "After infantile submission during rehearsals, there is a complete reversal when (the actor) is able to hold the audience in his grip and make them infantile and submissive."

Being able to control an audience is not enough to fill the identity gap in the actor's personality. The psychoanalyst notes that even stars have identity problems. They exhibit "narcissistic entitlement," a retaliation for the submission they have suffered. Unfortunately the actor who becomes a star, having so little sense of self, can only be a parody of the situation he suffered. He becomes a stereotype of a star, the only self he can assume. He "then becomes either perverse or reclusive."

Another powerful force influencing the morality and behavior of theatre people is the audience itself. The analyst source believes that audiences appreciate what the artist does because as children we all painted, danced, sang, and acted. Therefore, "we appreciate how hard it is to do these things well." He says that while everyone else has developed in a way that left the arts behind, the artist is a "mutant" for whom we have as yet no explanation. About whom there is no "developmental knowledge."

Because the audience remembers what it was like to be a child and an artist, it retains a nostalgia for the freedom of childhood. As adults we suffer through the demands of behaving responsibly. "In real life, all the virtues are corrupted by infantile aims." In other words, the audience roots for the actor who dares do things as an adult that the audience cannot. The actor can be an "exhibitionist without shame." The actor can commit crimes without punishment or guilt. In fact, there are times when the audience cheers on the villain.

Exhibitionism, narcissism, tantrums, willfulness, complicated sexual liaisons, all seem to be encouraged in the artist by a public that wants a vicarious revenge against the restrictions of the adult role. It is not hard to see how difficult it is for an artist to observe established moral strictures.

But it is too simple to say that the audience wants the artist to misbehave. The public also envies a star and loves to see his comeuppance. The tabloids thrive on bad news of famous people, stories about arrests, divorces, and scandals of all sorts. The troubles suffered by famous artists validate the rule we all learned while growing up: if you misbehave you get punished. The public is ambivalent. It wants to see the famous personality get away with some outrageous act and it also wants to see retribution. The public says, in effect, "We hate the rules we live by and wish we had the nerve to break them as you do, but if you are not punished then all the things we have been taught to believe about behavior are meaningless."

> *"The troubles suffered by famous artists validate the rule we all learned while growing up: if you misbehave you get punished."*

We have all been taught to think about others and not be self-involved, to sacrifice our own interests for the good of family, community, and nation. The artist, however, is unabashedly self-involved. How inconsistent it is that we who are taught to sacrifice are also shown example after example of the selfish behavior by geniuses whose histories we learn. Look at this incident from Francoise Gilot's book about her life with Picasso. On the day she went into labor, Picasso was scheduled to receive a prize, for his peace dove drawing. He insisted that the chauffeur drive him to his award ceremony before taking Gilot to the hospital.

While looking at the self-centered behavior of artists, we must take note of the increasing number of social critics who see narcissism becoming widespread in the general society. The *New York Times* of March 16, 1982, on new studies of narcissism:

> ...*therapists feel the new understanding of narcissism is giving them insights into the behavior of terrorists, actors, some leading businessmen, and even presidents.*

163

This interest in narcissistic disorders "coincides with certain features of our culture in our time," notes Dr. Kernberg, referring to such things as the premium put on one's ability to manipulate people and the reliance on appearance rather than substance.

If these observations are correct, they explain, in part, the increase of flash in theatre at the expense of content, the increased enrollments in our theatre departments, and the shrinking gulf between theatre people and the general public.

I do not believe that narcissism and artistic selfishness are the same. In searching through some recent articles about narcissism I find tremendous disagreement as to whether narcissism is always bad. Some feel it can be a healthy and productive force. Another view held by academic psychologists dismisses the term narcissism because narcissism cannot be measured scientifically. I am left, therefore, to my own devices, my own prejudices, my own completely unqualified assumptions about what I think are two different kinds of interest in self.

It seems to me that the narcissist in society is concerned with surface while the real artist is interested in substance, that the narcissist seeks pleasure and drifts from one new fad to another in search of gratification while the artist suffers a great deal of pain digging deeper and deeper in his search for what is inside himself. And if the artist is living in two worlds and is compelled to express his second one to the rest of us, he is driven by forces different from what psychologists say is the narcissist's reaction to an arrested development.

My prejudice in this matter does not stem from a desire to forgive artists their immoral acts but rather to dispel the notion that extreme self-involvement makes everyone an artist. Being an artist is difficult and often painful. Simply because artists often behave selfishly, it does not follow that a narcissist is an artist.

A director told me, "A great artist is not a normal person." And little wonder. He is at the same time adored and reviled by the public. He represents infantile freedom, but his vision is of value only inasmuch as it speaks to an adult understanding. Driven by ego and a need to express himself, he is cheated of money, made childishly dependent, and humiliated by those intended to help him achieve his goal. And in theatre, he must work cooperatively with other artists, each with the same monumental mass of contradictions and with individual visions to express. It is amazing that within these stormy forces there remains a modicum of ethical behavior. And yet there does.

I think that proper behavior in the theatre depends on two separate systems. For want of better terminology I shall label one moral and the other ethical. The moral code is philosophical or religious and concerns universal truths of good and evil that bind the larger world. The artist flouts these rules and may be rewarded by society for doing so. He who is notorious, whose sexual adventures are well-known, who disrespects authority, who is profligate with money—he draws crowds to theatres because of his eccentricities. The ethical code is a group rule that governs those necessary working regulations that make performance possible and somewhat efficient. An actor's colleagues do not care (except as gossip) whom he is sleeping with, what drugs he is buying or selling, or what terrible things he has done to his mother, so long as he arrives at the theatre on time, remembers his lines, doesn't upstage his fellows, and doesn't steal scenes.

Difficulties may result from an artist blurring the lines between moral and ethical conduct. He who is feted, spoiled, and in demand for his outrageous public behavior may find it hard to conform to group professional rules. If he is powerful enough and needed enough he may get away with ignoring theatre ethics. But his colleagues will complain among themselves and easily recognize and condemn a breach of their code.

I asked many of the people I interviewed if it was possible to be immoral and still be a great artist. This undergraduate conundrum was answered by sophomores years ago. Some said that it was impossible to be a great artist and a bad person at the same time. Others said that it was possible for a short time—then immorality would begin to affect art. At the opposite extreme, one man said, "Greatness and bad behavior often go together. You can be evil, a terrible person, and yet act a saint on stage."

My own sense of fairness makes me prefer the idea that immoral artists are bad artists; experience forces me to admit that the opposite is true. None of this is terribly important except that it returns us to the first objections to theatre: theatre portrays the good as well as the evil, theatre leads man away from the truth. This criticism is voiced from Plato to the present. In his recent book, *The Antitheatrical Prejudice*, Jonas Barish says,

> *Human existence can hardly avoid resembling in basic ways the experience of actors in the theatre, and human consciousness can hardly escape the tinge of bad faith this introduces into our actions, the incitement it gives us to wish to be admired, stared at, made much of, attended to. By living in the theatre, as they do, by giving themselves over to mimicry and exhibitionism, men jeopardize the most precious part of their humanness, the right exercise of which would be a relentless campaign to rid themselves of the element of falsity in which they move.*

So here we are 2,500 years after Plato, still damned for an occupation that society finds necessary. Like whores, hangmen, and money lenders, theatre people are a tolerated group, allowed to serve the world at the risk of their own souls. While the play as an end product may further the search for truth, the players themselves are condemned.

And yet, the exclusion from society of the theatre artist, of all artists, may be the only way to truth. Childish, self-centered, with a vision of other worlds, the artist is freed from the constraints of society and, more than anyone else, allowed to speak the truth as he sees it. It is the artist's anomalous state that makes him valuable.

A few years ago I met Jozef Szajna in Warsaw. What he produced and his means of producing it seemed at odds and pointed out what, to me, is a constant and serious question of morality in theatre. An official Polish brochure describes Szajna's production of *Replique* as "...the Apocalypse of civilization, chaos, tensions and fear in which one searches for truth, hope and faith in man." The brochure goes on to explain the source of his vision: "Many of his spectacles are full of obsessive reminiscences of the horrors of the concentration camp, which he personally lived through as a young boy in Auschwitz." Szajna does indeed create powerful theatrical images which search for truth. He is a great artist. But he controls his actors as if he is a guard in a concentration camp and they are prisoners. At one point in our conversation he said that if an actor had an idea about the production (the plays are non-verbal and created like choreography—improvised, added to) then he, Szajna, would do exactly the opposite thing to keep the actor off balance. His actors are the tools of Szajna's personal vision. No matter how glorious that vision is, the actors remain merely tools.

Szajna is one extreme. The other extreme would be a democratic production in which every participant—actors, designers, director—would have a voice and negotiate their vision of the final product. Most productions in our theatre fall in between with a director listening to suggestions but having the final say. It is within this spectrum of working relationships that the collaborative system of theatre is constantly tested—for there is no established formula. In the process of testing, the participants use all the devices we have been looking at: power, emotional blackmail, persuasion, money, ego, humiliation, greed, and often sex. All of these are used at one time or another to allow one individual to express his vision at the expense of another's.

I am not suggesting that every production is a seething pool of horrendous relationships. When people have worked together and mutual trust and understanding have grown through a continuing association, the chance is greater for a harmonious production.

Sometimes personalities just seem to hit it off from the beginning and no problems follow. But I have been discussing problems. Some people think that every production presents difficulties. Which reminds me of Boris Aronson's maxim for theatre which was quoted by several of those I interviewed. They invariably imitated Aronson's accent: "Are two rules in theatre. Rule one: in every production is a wictim. Rule two: do not be da wictim."

Could it be that the underlying conflict in all the moral and ethical questions that disturb us in theatre is that we cannot avoid using others as means to an end? The nature of a collaborative performance art is such that we need others to fulfill our own visions, to support our own egos, or to fill our own pocketbooks. This in itself goes contrary to an adult non-narcissistic concept of morality. H. L. Blackham in *Moral Theory and Moral Education,* writes, "That every human being is an end in himself, and not merely a means to an other's ends; that every human being is a subject like oneself, and not merely an object to oneself; this is the underlying rule of coexistence and cooperation which justifies the practical rules which regulate society."

To adhere to Blackman's principle one must be adult, understanding, and very secure. This holds true in the theatre as well as in the wider society where we find plenty of vicious, unfeeling behavior. But in theatre, we seem to deal with infantile ego needs, frustration, and fear more than does the rest of the world. We attract the misfits who have larger needs and larger dreams, and personalities with kinks. We appear more able to suffer the aberrant vision and we reward childish exhibitionism and make believe. And like war, we depend on end results to justify the means.

All of these things may make it impossible for theatre to ever be an activity where adult morality holds sway. But it may be that our imperfections serve society after all. We can use our corrupt tools (power, greed, selfishness, and humiliation) to fashion those theatrical dreams that lift, free, and inspire an audience. Our ability illuminate the human condition while under the shadows of our own perversity is the paradox of theatre.

A Letter To A Young Designer

First published in **Lighting Dimensions**: June 1983.

Dear Friend,

I watched you at work in the auditorium, your drawings spread over the wide desk and the shaded lamp lighting the changes in your expression. The intercom was in your one hand while your other hand searched over lists of numbers and symbols on your plot. People were sitting all around you talking. Other voices came over your intercom; voices came from the stage, all of them urging, begging, commanding, flattering, damning. You were a lightning rod for all the psychological electricity unloosed on the final days before opening.

We had talked about what might happen to you on this, your first professional job. But nothing could truly prepare you. So many variables are at work, including your own reactions under pressure. There are no scripts, no road maps, no standard procedures to follow. As I suspected, you found your own way, even though at times you doubted you would survive. You did survive, even triumphed. Your lighting was close to what you wanted it to be and the experience made you a veteran, slightly battle scarred, but confident in your ability, stamina, and purpose.

Those last few days of rehearsal are completely different for the lighting designer than for anyone else. He is the sprinter on the team. He must summon up one fast concerted burst of energy to do his job. Or maybe he is like the last runner on a relay team. By the time the others have passed him the baton, they have done their best and look to him to win or lose the whole race. The pressure is on him, regardless of how slow the others have been.

As you worked in the crowd, you were envious of the set and costume designers who worked alone in their studios, no one looking over their shoulders until they had made their work what they wanted it to be. Sure, you had a bit of time to focus and set a few levels, but after what seemed like minutes the army of

Genghis Khan descended with demands for space and time to practice their barbaric rites. They sang and danced, moved sets around, and shouted through the once dark and calm theatre.

Until then you were working with equipment, steel and glass, cables and lamps. When the hordes arrived you also had to deal with egos, personalities, prejudices, and insecurities. You expect to make this abrupt switch in your attention, but there are always surprises. No matter how firm the script of the show, how perfect the technical production, how finished the lighting design, the lighting designer is always working in improvisational theatre.

It was also at this time that the others looked to you to be a god. A lighting designer does appear to have supernatural powers. His control over arcane technical equipment, combined with the visible magic he performs, give him an aura of influence and at the same time creates expectations that are hard to satisfy.

"I'd like that girl levitated about three feet. Can't you do it with light?

At other times you were expected to be a janitor, a kind of artistic garbage man, carting away the mess that others have made.

"That red sofa should disappear in this scene. Can't you light it out?"

It is amazing how some in theatre think of light as being able to hide things. Part of this goes back to the magic you do perform, but many times there is a real ignorance about the nature and use of light. And there is no time in the frantic hours of rehearsal to educate a director or producer. When you say you can't hide this or that or make a huge object seem small, they blame you; they are disappointed and frustrated.

Frustration! It is the motor that drives every sort of aberrant behavior in the final days of rehearsal. It is the Eumenides, never invited but always present to needle, push, and betray one's best instincts and moral resolves. The interesting thing is not your own frustration; that can be defined in terms of your goals and how far you are from them. Nor is it the collective frustration, one that joins the whole company in a shared concern for a flawed production. The fascinating thing to watch is the way the individual frustrations grate against each other causing friction, heat, and pain.

Young designers, like you, are often unaware that everyone else, or should I say the real artists working on a production, have fallen short of what they envisioned their work to be. It is never as good or as beautiful as what their mind's eye saw in those quiet moments of preparation. The young designer is blinded by a myriad of his own problems and can miss the pain and frustration that others feel. He only sees the outward reactions to frustration: anger, nastiness, the dissatisfaction with the work of others.

Directors are especially prone to fits of frustration. Theirs is the central vision, yet they have no direct control over anything. They have no concrete tools but must rely on intelligence, persuasion, strength of character. In those last rehearsals, the one thing a director might actually control is the lighting. There it is being worked on in front of him. It seems that all one must do to effect changes is to say a few words into a headset. There you are, the lighting designer, prey to the frustrations of the director. It is too late to change the acting, or the script, or the sets. What he sees on stage is not what he dreamed of seeing. You, then, are his last chance of doing something to bring his show closer to his dream. At this point doing anything is better than doing nothing.

There are, of course, variations of the frustrated director syndrome. The show is so far gone that the director has given up and does not bother you. The director is not an artist so he is not frustrated. The leading man can't remember his lines so the director spends his time working with him. The director is technically ignorant and visually blind and is afraid to mess with the production. In some ways it is therefore better to have a director badgering, screaming, and cursing because it is a sign that the show is close enough to fight for and there are no larger problems.

I know that you would like to have advice on ways to handle each situation. My frustration is that I cannot tell you how. You will deal with things in one way while others cope in their own ways. Each one finds a combination of wheedling, cajoling, lying, tricking, and a good dose of being so efficient, talented, and right that one is beyond criticism. I saw you in these rehearsals and you did well. OK, there was that one late night blow-up when everyone was exhausted. But you and the director are still friends. That means a lot.

You are on your way. Godspeed.

"...THE BEST IS YET TO COME."

First published in **Lighting Dimensions**: September/October 1981.

Mozart was approached by a young man who asked him how to write a symphony. The composer suggested that the man start by writing songs and chamber music.

"But Maestro, must I begin with such simple exercises? You wrote symphonies when you were nine."

"Yes. but I didn't have to ask how."

• • •

In a recent article, I wrote about designers who teach but miss many classes because they are busy with professional work. In response to this piece, an administrator of a theatre department told me he thinks one reason why professional/academic designers set themselves such a busy schedule is that design is a young man's game. Designers know that if they don't work while they can, younger designers competing for jobs will rob them of their careers. This idea was proposed by an administrator who has worked with young and established designers. He is a man of knowledge and perception. His theory deserves examination.

In 1970, Simone de Beauvoir wrote an exhaustive (and exhausting) work *The Coming of Age*, about the effects of age in relation to biology, ethnology, history, everyday life, and, what concerns us here, the arts and sciences. She attempts to identify those periods of life during which men and women, in a variety of professions, do their best work.

For example, the scientist and mathematician do not produce new ideas past their middle years. A middle-aged scientist has already made his contribution. There are always exceptions: Galileo, Franklin, Michaelson. For most others, breakthroughs come after the basic knowledge has been mastered but has not hardened into unquestionable rules. The young scien-

tist sees new possibilities and can fearlessly look ahead to the years he will need to prove his ideas.

After the scientist has made his mark, his self-interest dictates that he defend what he knows against the pressure of new ideas he doesn't understand. Beauvoir quotes the French philosopher of science Gaston Bachelard, "The great scientists are useful to science in the first half of their lives and harmful in the second." Look at Edison, who at age thirty-one invented the electric light, but later in his life opposed the introduction of alternating current.

Writers, according to Beauvoir, have a similar early flowering—but for different reasons. The writer has a love-hate relationship with the world. He wants to be noticed and assigns great importance to the world even as he rejects the way things are. The resulting tension between things as they are and the way an author would have them be is the driving force of his work and imagination. Men grow more accepting as they grow older and thus lose the passion and strength necessary for fiction.

Berenson is quoted as saying, "What a man writes after sixty is worth little more than tea continually remade from the same leaves."

Again Beauvoir cites exceptions. "Sophocles produced *Oedipus at Colonus* when he was eighty-nine. Voltaire wrote his finest works in the last twenty years of his life." But if a writer continues to be productive in his later years, it is often not through fiction, but with essays or memoirs, works that draw less on inner turmoil and more on actuality. Also, the fear of repeating previous ideas in watered-down fictional form is alleviated by describing real events.

Beauvoir finds that musicians improve with age. "In order not only to do new things within the set rules but also to break free from them to a certain extent (the musician) needs a great deal of self-confidence, and therefore a considerable body of work behind him."

In much the same manner, painters spend a lifetime mastering their craft and gaining self-confidence. At the end of their lives they are able to ignore the restraints of public opinion, while at the same time (unlike novelists) they are constantly bombarded with external inspiration. As Beauvoir says, "...there is always some thing to paint." As opposed to the scientist, a painter facing a new project looks ahead to its completion in hours, days, or weeks. The discouraging prospect of work stretching into years and ultimately remaining unfinished is not part of the painter's experience and saves him from giving up at the end of his life.

It has always seemed to me that designers, like painters, are late bloomers, doing their best work in later years. I checked with

Domingo Rodriguez, business representative for United Scenic Artists. He sees every union contract in New York and knows exactly who is designing what. His knowledge confirmed my prejudice. Most of the work is being done by designers who have been around for a while and have amassed a store of experience. A short list would include Robin Wagner, David Mitchell, Eldon Elder, Bill Ritman, Tony Walton, and Ming Cho Lee. It is exceptional when an Andrew Jackness or a John Lee Beatty gets an opportunity to design a big show in New York.

There are reasons why a talented young designer has a hard time. There is that Catch-22 situation where the producer wants to trust his million dollar show to someone with a track record. This keeps many young designers out in the cold wondering how to get that first break. But since there is an increased authority and competence that accompanies a growing body of work, a producer is not arbitrary in hiring experience; he will, according to this theory, get a better show.

There is evidence that designers do their best work in later years. The most obvious example is Boris Aronson who would have been eighty-one this year. His earlier work was fine, but his brilliance shone more brightly in the string of musicals he did in his late years: *Company*, *Follies*, etc. They were inventive and not the work of an old man repeating himself.

Some designers begin to soar at mid-career. Certainly Jo Mielziner took off like a rocket at this age. When he was in his late forties he designed his most notable works: *Death of a Salesman*, *A Streetcar Named Desire*, *The Glass Menagerie*. And he went on in the next years to do *South Pacific*, *Guys and Dolls* (in the same year, 1950), and *The King and I* (the next year).

A look at recent "Oscar" and "Tony" winners shows that their ages are from the late forties to over seventy. I include here John Bury, Tony Walton, Phil Rosenberg, George Jenkins, and David Mitchell.

A list of the most successful lighting designers might also show a correlation between age, experience, and employment; however, because so many of them are women, I will be both sexist and gentlemanly by not mentioning names or ages.

It is obvious why experience, seeing what works and what doesn't, plays an important part in a designer's development. Another factor that keeps older designers growing is that the major inspiration for a designer is external. Interpreting verbal ideas in visual forms relieves the designer of the need to constantly dredge his artistic pool. I could extend the metaphor to say that a designer's pool is constantly fed by the springs of a playwright's ideas.

All of this may be discouraging to young designers who naturally want success in a hurry. It should not be. Getting better at what one does throughout a lifetime has a great benefit. It nurtures hope. What can a dancer or an athlete look forward to?

If anyone needed a reason to look ahead with hope, the cases of Frank Lloyd Wright provides it. According to Tom Wolfe's "From Bauhaus to Our House" in the June and July *Harper's,* in 1935 Wright designed "Falling Water," the famous house cantilevered over a waterfall. He was then sixty-nine years old. Between that year and when he died in 1957, at ninety-one, he produced 180 buildings, more than half of his life's work.

I can hardly wait to grow old.

The Last Interview Or
Ask Me No Questions And I'll...

· ●

First published in *Lighting Dimensions*: April 1982.

*L*ighting Dimensions recently paid a visit to the studio of Pericles MacCohen where Mr. MacCohen graciously answered some of our questions. MacCohen is considered one of this country's top 11,364 designers. Now at the age of ninety-seven MacCohen granted us his first interview in ten years.

LD: Mr. MacCohen, why has it been so long since you last gave an interview?

MacCohen: No one asked for one.

LD: But surely people are interested in your work You have done so much. How many shows have you designed?

MacCohen: I've designed 627 Broadway shows, 341 off-Broadway shows, 212 ballets, 96 operas, 12 circuses, 23 restaurants, 62 feature films, 422 commercials, and 68 porno films.

LD: Wow!

MacCohen: But I didn't do sets, lights and costumes for all of them. When I started there wasn't much lighting and the porno film costumes were only black socks and masks. For the rest, I always wanted complete design control. That's what's wrong with the young folks today, no control. They're soft, mushy, lousy kids. No idea of what they are doing. Shouldn't be allowed behind a pencil.

LD: You sound a touch bitter.

MacCohen: You're dang right I am. Pimply faced sixty year olds are taking over the business. Just when you get good in your seventies and eighties those whippersnappers underprice you. I guess this is just a young man's business.

LD: What advice would you give a young designer?

MacCohen: Go into insurance or used cars. Don't make yourself crazy with theatre. If they were smarter I'd suggest law or medicine but these kids are dummies.

LD: But many young designers don't want to do anything else. They love theatre.

MacCohen: When I was sixteen I got a job as a marriage counselor but that didn't work out. I tried lots of other things until I got a job as a salesman of cello bows. That took me into lots of theatres. You know I had to see every cellist on the circuit. I hung around making myself useful and one day the scenic artist was so drunk he couldn't finish the oleo drop they needed for the evening performance. Since there were hairs on cello bows and hairs in brushes they thought I could finish the drop.

LD: And it was such a success you became a painter?

MacCohen: It was so terrible they threw me out of the theatre. And I was mad. No one could treat me that way. So I practiced up. I was going to get so good at painting that one day I would walk into that place as a master painter and they would beg me to paint for them.

LD: And what happened when you finally went back?

MacCohen: I never did. The theatre burned down two weeks later.

LD: How did you break into theatre?

MacCohen: Painting large drops is hard when you have no drops to practice on, so I painted small. You know the tissue paper they pack cello bows in? I used the tissue paper to draw all the scenes they used for drops: park scenes, woods, city streets. One day I was delivering a bow and Tarmac Shlumsky saw one of my drawings when it fell out of the bow box. He bought six drawings and I was a designer. See what I mean? How can you tell kids to sell cello bows if they want to break into theatre?

LD: You have had a lot of successes. What are some of your favorite sets?

MacCohen: Most people ask me about *Who's Afraid of the Kitchen Sink* because there were so many things on the set. A junk yard of a set. A sewer. A rat's nest of trash. It won a Tony and a grant from the Environmental Protection Agency. Let me tell you about that one. The set I designed had a nice linoleum floor but it was curling up. The carpenter glued it down and then piled junk on it to hold it in place. The director saw it and liked the junk "Add more," he said. Nothing was enough. "Add more, add more." Under all that stuff was the set I had designed. Then there was *Goys and Dills*. A very popular show. And you know why? Because of the underlying theme I gave to everything. Up till then, no one had used pickle colors in a show. I used every shade of pickle color. It gave the show a real flavor.

LD: That was a wonderful idea. How did you think of that?

MacCohen: I didn't think of it. I was called away unexpectedly on Friday and my lunch was on the drawings all weekend.

LD: You mean...?

MacCohen: Sure, the pickles colored all the sketches. But the most interesting color combination was for *Drizzle*. My assistant, Alvin, was sick to his stomach...

LD: I think we'll skip that story. Mr. MacCohen, since this is a lighting magazine, can you tell us about your theories and experiences lighting shows?

MacCohen: There is only one thing to remember, the number three—hit 'em three ways at once and always hang three times the lights you think you'll need.

LD: But isn't that expensive?

MacCohen: Not if you get results. Take the ballet lighting that everyone uses now. That was my invention. Shin busters were the result of a loaded boom. There were so many lights on the wing booms that they went from top to floor. When the bottom ones were turned on, POW. A great effect. And then on another show I invented the electric pipe in view of the audience.

LD: What aesthetic theory did you use?

MacCohen: Theory, schmery, the pipe was so loaded that the crew gave out before they could get it higher. Again it was a case of the director liking it. "I like it, I like it," he kept shouting. The idiot. "Keep it in," he says.

LD: There must be some theories or rules that our readers would like to hear about.

MacCohen: There is one good lighting rule I follow and it always pleases an audience. Lots of twinkly lights. Little lights, big lights, practicals, sparklers on drops, on portals, chasers, gas lamps, candles, chandeliers.

Twinkle twinkle little stage
Sparkle lights are all the rage

They clap their filthy hands off for strings of lights. Dummies.

LD: Are you saying that cheap effects are popular?

MacCohen: Have you been listening to me? You're a dummy, too. Idiots all around me. Look—put some running water on stage, drop in an American flag, have twenty bimbos kicking up their heels in unison, have a moonlight scene with stars, use a turntable and all the dummies go mad. They think you're a genius. I'm giving you the real stuff, fathead.

LD: You are being quite candid. Can I ask you then, the secret of your success?

MacCohen: You sure ask stupid questions. Don't you guys ever have any new questions? Look half-wit, the secret of anyone's success is luck. It's luck, luck and more luck.

LD: Are you suggesting that things happen by chance?

MacCohen: Only the best things, simp.

LD: And ordinary things?

MacCohen: They are designed. Get out of here, Bozo, out, OUT.